High Stakes, High Hopes

GEOGRAPHIES OF JUSTICE AND SOCIAL TRANSFORMATION

SERIES EDITORS

Mathew Coleman, *Ohio State University*
Ishan Ashutosh, *Indiana University Bloomington*

FOUNDING EDITOR

Nik Heynen, *University of Georgia*

ADVISORY BOARD

Deborah Cowen, *University of Toronto*
Zeynep Gambetti, *Boğaziçi University*
Geoff Mann, *Simon Fraser University*
James McCarthy, *Clark University*
Beverley Mullings, *Queen's University*
Harvey Neo, *Singapore University of Technology and Design*
Geraldine Pratt, *University of British Columbia*
Ananya Roy, *University of California, Los Angeles*
Michael Watts, *University of California, Berkeley*
Ruth Wilson Gilmore, *CUNY Graduate Center*
Jamie Winders, *Syracuse University*
Melissa W. Wright, *Pennsylvania State University*
Brenda S. A. Yeoh, *National University of Singapore*

High Stakes, High Hopes

URBAN THEORIZING IN PARTNERSHIP

SOPHIE OLDFIELD

THE UNIVERSITY OF GEORGIA PRESS
Athens

Publication of this open monograph was the result of Cornell University's participation in TOME (Toward an Open Monograph Ecosystem), a collaboration of the Association of American Universities, the Association of University Presses, and the Association of Research Libraries. TOME aims to expand the reach of long-form humanities and social science scholarship, including digital scholarship. Additionally, the program looks to ensure the sustainability of university press monograph publishing by supporting the highest quality scholarship and promoting a new ecology of scholarly publishing in which authors' institutions bear the publication costs.

Funding from Cornell University made it possible to open this publication to the world. www.openmonographs.org.

Further support for an Open Access edition was provided
by the University of Basel, Switzerland,
and the University of Cape Town, South Africa.

© 2023 by the University of Georgia Press
Athens, Georgia 30602
www.ugapress.org
All rights reserved
Set in 10.25/13.5 Minion Pro 3 Regular by Kaelin Chappell Broaddus

Most University of Georgia Press titles are
available from popular e-book vendors.

Printed digitally

Library of Congress Cataloging-in-Publication Data

Names: Oldfield, Sophie, author.
Title: High stakes, high hopes : urban theorizing in partnership / Sophie Oldfield.
Other titles: Geographies of justice and social transformation ; 60.
Description: Athens : The University of Georgia Press, 2023. | Series: Geographies of justice and social transformation; 60 | "Publication of this open monograph was the result of Cornell University's participation in TOME (Toward an Open Monograph Ecosystem), a collaboration of the Association of American Universities, the Association of University Presses, and the Association of Research Libraries." | Includes bibliographical references and index.
Identifiers: LCCN 2023003901 | ISBN 9780820365008 (hardback) | ISBN 9780820365015 (paperback) | ISBN 9780820365022 (epub) | ISBN 9780820365039 (pdf) | ISBN 9780820365046
Subjects: LCSH: Cities and towns—Research—South Africa—Cape Town. | Housing—South Africa—Cape Town. | Community and college—South Africa—Cape Town.
Classification: LCC HT110 .O434 2023 | DDC 307.76072068735—dc23/eng/20230130
LC record available at https://lccn.loc.gov/2023003901

CONTENTS

LIST OF LITERATURE BOXES vii

LIST OF FIGURES ix

ACKNOWLEDGMENTS xi

PROLOGUE 1

CHAPTER 1. **In Partnership** 3
The Partnership's Cast of Characters 5
The Book's Design 7

CHAPTER 2. **A City, a Community, a University, a Partnership** 10
Staying, Not Running 10
A Partnership 14
A Turn to Narrative 18
Making Visible the Partnership and Its Practices 21
Ordinary Words, Urban Worlds 23
Coda—A Partnership 25

CHAPTER 3. **A Decade: A Chronology of Projects** 26
"We Need Help" 26
The Civic, an Introduction 27
Getting Started, a Project on Backyarders 34
Out of the Classroom, into the City 35
Land Occupation, a Shift in Research Agenda 38
Who Has Moved into Agste Laan? 42
Civic Work and Its Wide Parameters 44
Minstrels as Community Development 47
Making Ends Meet, Neighborhood Economies 52
Incremental Rhythms 56
Coda—A Process 58

CHAPTER 4. **Crisscrossing Contradictions, Compromises, and Complicities   60**
Navigating   60
Ek Is Die Baas!   61
Don't Worry Lady, I Have a Gun, I'll Shoot!   66
Disquieting Differences in a Wilted, Waterless Garden   68
A Partner, a Land Invader, a Ward Forum Member   70
Fear, the Complicities of Xenophobia   71
A Complaint   74
Didn't You Wonder Why? Neighborhood Crime and Violence   75
An Endpoint   77
An Empty Fridge   79
Coda—Contradiction   79

CHAPTER 5. **Teaching and Learning: Across the City, Back and Forth   80**
Onto the Bus   80
A Gangster Snap, a Zoo   84
Engage with Respect, a Guide   88
Questioning What We Know   91
In Homes, Not Shacks! Interrogating Readings   95
A Toolbox for Writing   97
Critique Leavened with Love   101
Coda—Teaching   106

CHAPTER 6. **Research: A Web of Writing Practices and Publics   107**
Writing Practices   107
Research Posters Taped to Walls   108
"There We Are on the Map"   112
Located in Journal Articles   116
Yellow Pages in Every Household   119
"That's My Book!"   122
"The Story of Sewende Laan Is like a Book"   126
Spinning Off, Student Research   130
An Archive across the City   132
An Interwoven Web   134
Coda—Publications   135

CHAPTER 7. **Theorizing the City Otherwise   136**
In Stories of Collaboration   136
In Ordinary Words   138
In Verbs—In the "Doing Words" of Practice   141
Theorizing in Partnership   144

IN MEMORY OF GERTRUDE SQUARE   147

BIBLIOGRAPHY   149

INDEX   155

LITERATURE BOXES

A South African Imperative: Urban Studies "Otherwise"   11
Refiguring: Collaboration and Its Inspirations   17
Experimenting: Writing Urban Studies "Otherwise"   20
An Invitation: Urban Theorizing "Otherwise"   24
Doing Urban Studies "Otherwise"   27
Struggles for Housing   37
Activism as Participation   46
Klopse and Its Layers of Politics   51
Struggles to Make Ends Meet   53
Between Refusals and Invitations   61

FIGURES

1. Valhalla Park, an introduction   12
2. Civic leaders, partners   28
3. Between the Civic and the city   32
4. Mapping Sewende Laan   40
5. Building Agste Laan   43
6. What it takes to lead   45
7. Competition time for Valhalla Park minstrels   48
8. Making ends meet   54
9. Teams of researchers   82
10. Running and recording fieldwork   85
11. Interviewing together   89
12. Guidelines for Research   90
13. Reams of journals and reports   98
14. Contrasting assessments   104
15. Research posters   110
16. The Agste Laan map on the wall   114
17. Deciphering the practice   118
18. Our *Yellow Pages*   120
19. The Klopse book   124
20. *My Sewende Laan*   128
21. Scrutinizing the posters   133
22. Gerty and Sophie, in the early years   147

ACKNOWLEDGMENTS

I am indebted to many wonderful people who were part of the partnership, its building and running, and the writing of this book. Heartfelt thanks to my partner the late Gertrude Square. Her wisdom, courage, and heart inspired me and guided me in profound ways. I miss her deeply. Washiela Arendse and George Rosenberg, both sadly now deceased, welcomed me into Valhalla Park; their commitment across the years was a foundation for this work. A special thank you to the Square family. Many of you were part of the partnership and have become my family in so many ways: Zaida, Miena, Shireen, Kader, Leaticia, and your own families. It has been a privilege to work with long-standing research partners: Dan, Fatima, Koekie, Masnoena, Rosemary, Suki, Lefien, Naomi, Faranaaz, Aunty Meisie, and Aunty Fadielah, Antony, Nawaal, Sylvia, Jamiela, and Eric. You have taught me so much and inspired and sustained our partnership.

I am fortunate to have worked with exceptional and generous colleagues. They have been formative to my work and this book and its completion. Brenda Cooper gave me the courage to write stories and had faith in my capacity to do so from beginning to end. Richa Nagar has been a constant inspiration, critical interlocuter, and a dear friend. The late Elaine Salo was my partner in "crime." I remember and treasure our teaching and writing together, her joyous laughter and incisive critique. Anna Selmeczi, Antonádia Borges, and Claire Bénit-Gbaffou have been precious colleagues, the dearest of writing friends, brilliant in care and commentary. Lucia Thesen's conversation and wisdom helped nurture and bring the book to fruition, especially over our shared fence under lockdown. Alma Viviers offered her visual creativity and skill, shaping a layer of this narrative that I had not imagined. Laura Nkula-Wenz and Myriam Houssay-Holzschuch, and the Menthonnex team, cheered me on and set the bar for writing collegiality. The now late Clive Barnett gen-xii Acknowledgments

erously read and patiently commented on an early draft. Tanja Winkler and Koni Benson inspired me and kept me going over the years. Saskia Greyling was there from the beginning (as a student) to the literal end, a colleague and friend engaging in this book in such meaningful ways.

This work grew and developed over the long haul in the Department of Environmental and Geographical Science at the University of Cape Town (UCT), and in its later stages at the African Centre for Cities at the University of Cape Town, and the urban studies section in the Department of Social Sciences at the University of Basel. Thank you especially to colleagues and friends Susan Parnell, Sharon Adams, Maano Ramutsindela, Michael Meadows, Jane Battersby-Lennard, Shari Daya, Vanessa Watson, Edgar Pieterse, Henrietta Nyamnjoh, Bradley Rink, Janice McMillan, Suellen Shay, Sonwabo Ngcelwane, Shirley Pendlebury, Robert Morrell, Alan Mabin, Francis Nyamnjoh, Divine Fuh, Tim Stanton, Maren Larsen, Kenny Cupers, Jinty Jackson, Geetika Anand, Marcelo Rosa, Dolly Mdzanga, and Neema Kudva. I have also had the privilege to work with Joanne Bolton, Robyn Rorke, Siân Butcher, Saskia Greyling, Inge Salo, and Raksha Ramdeo-Authar, fantastic students who played key roles as assistants in the juggling act of working with me to make the partnership tick. It has been a joy and privilege to work with each of you.

Students from the University of Cape Town, and in later years from Stanford University, threw their hearts into the partnership and its projects. This work has been a teaching test bed, and I treasure the ways in which it became an inspiration for collaborative research studio work with a wonderful next generation of masters of Southern Urbanism and masters of Critical Urbanisms students. This City Research Studio work was designed and run with Noah Schermbrucker, Dolly Mdzanga, and Shawn Cuff at the nongovernmental organization Peoples Environmental Planning; with community-based partners Lele Kakana in Napier in the Overberg in the Western Cape, and Charlotte Adams in Hazeldean-Ekupumleni in Cape Town; and with Adnaan Hendricks and Melanie Johnson in Ruo Emoh in Cape Town.

I had the opportunity to develop this book with the generous support of the Department of Environmental and Geographical Science at the University of Cape Town, a Mandela Fellowship at the W. E. B. Du Bois Institute at Harvard University, a Programme for the Enhancement of Research Fellowship at the University of Cape Town, a Community Engagement South African National Research Foundation Grant, a British Academy Newton Advanced Fellowship, and funding from the University of Basel. Without this financial and intellectual support, this publication would not have been possible.

Thank you to my family and friends who have listened across so many years and supported me in so many ways: my sister, Bronya Oldfield, my parents, Julienne and John Oldfield, my hiking friends who walked and talked the partnership and this book, Amy Mulaudzi, Tanya Jacobs, Anne Magege, Tanja Winkler, and Tania de Waal. Most of all thank you to David and Zoe Maralack, whose love and encouragement, and patience, are at the heart of it all.

High Stakes, High Hopes

PROLOGUE

I run, rhythmically, one foot forward,
step after step, breath in, breath out.
The tar underfoot is stippled, torn,
a shattered glass shard, trampled,
A bottle top, squashed.

Dear Aunty Sophie . . .
Is this letter a dream or a nightmare?
A compass caught in a web; through it might I find my way?
The rhythm of an archive, a ringing bell, a call to prayer.

The words swirl and land,
pocket sized, powerful, potent, bespoke.
I think of my mother at my father's funeral.
She asks us to listen,
to not worry about exact meaning, precise words.
She is reading it for him, for her, for them.

A single solitary artichoke heart in flower
sits atop his coffin.
The curtain flutters, it moves in the slight breeze.
It shimmers, orange, then green,
carefully manufactured patterns woven in its threads.
The words wash over us.

CHAPTER 1

In Partnership

High Stakes, High Hopes builds urban theorizing in partnership, between a township neighborhood grappling with the legacies of apartheid, the neighborhood's community organization (its "Civic"), and the university tasked to research and teach the city. This theorizing emerges within the political and physical realities of everyday life. The rhythm of this book—and its theoretical argument—unfolds in stories. Its aesthetic form pulses with narratives, which share the logics and rhythms of the partnership:

- in a city bursting at its seams, struggling to deliver services, to manage the conflicts that threaten to tear it asunder
- in a township neighborhood grappling with evictions, forced to fight for every right, service, and resource
- in a university, high up on the mountain slopes, whose mission is to theorize the city's pasts and futures and whose legitimacy to do so is contested

The purpose of the partnership was to teach and research the city together. Over the course of a decade, neighborhood partners and I experimented to build the research process and pedagogy. The motivation for the approach we crafted was political and urgent. Through it, the partnership engaged durable, intractable neighborhood challenges and conflicts, the rapidity and violence of city change and its shifting geopolitics. Our partnership immersed us in everyday urban realities that shaped the demands of activism, the hardships of structural inequality, and the struggles for a right to the city. The partnership enmeshed us in the complex challenges that shaped the neighborhood—its racialization and segregated history and present.

In its collaborative method and pedagogy, our approach rooted teaching and research in the struggles of the neighborhood, embodied in the literal

Chapter One

and epistemic violence that entangled and divided both it and the university. Our approach was inspired by the neighborhood, in the often-contentious organizing and mobilizing by the neighborhood Civic, and by the practices that neighborhood residents enacted to survive, to make do, to live fully. The partnership was a means to work together collaboratively to engage and understand—to teach and research—these realities. Through the partnership, teaching was grounded in the city, in its racialized inequities, its materialities and creativities. It infused the city and ordinary people into the classroom. Through this collaboration, we engaged the perspective of ordinary residents, the activism of the Civic, its location and positioning in the city.

An always productively compromised collaboration, our partnership stretched us, extending university notions of critique and truth. We reworked conventional academic practice, reshaping the nature of critique. Doing so allowed us to forge a space for creative methodologies and epistemologies, for ways of knowing together. The partnership offered collaboration with substance. Through it, partners, students, and I reflected critically on epistemological questions: how we produced knowledge, with whom, and for what varied and multiple agendas.

The analysis unfolds in narratives built on ordinary words and acts. The way partners and I worked together became substance, the inspiration for our questions. Ordinary words, and the stories in which they emerged, show the genealogies of our partnership practice and its process. They share the ever-extending and always partial ways in which we came to know and work together. These collaborative practices encompass the conceptual tools that enabled our theorizing together.

In its narrative form, *High Stakes, High Hopes* shares the ways the partnership proved a vehicle for neighborhood partners, my students, and me to travel back and forth across the city and between the campus and the neighborhood. The stories in this book show ways in which we (partners, my students, and I) navigated thinking and theorizing in these spaces and through these relationships. The partnership shaped our questions. Teaching and research deepened and my writing shifted, immersed in these multiple publics and political questions. Provocations and challenges created a collaborative form and praxis. We produced an archive, a celebration of diverse ways to know the city, a celebration of urban research. The construction of this book— its aesthetic form—invites your engagement as a reader. This fuller accounting aims to bring the reader into the complexity of choice, of context, of thinking, of doing.

Through this narrative analysis, I consider a set of critical questions for urban scholarship. I examine the ways in which collaborative partnerships open provocative conversations on everyday urbanism: what it takes to sustain households in overcrowded homes and shack settlements; the realities that shape the demands of activism; and, finally, the hardships of structural inequality that intertwine in this neighborhood and city. I track ways the partnership was built, incrementally, through pedagogies to work and teach together, to research, to write, and to share our thinking in and across the urban inequalities that divided us. I reflect on what was at stake in the partnership and its creative, and at times conflictive, evolution. What changed in learning when teaching and assessment moved in and between the university classroom and township streets and ordinary people's households? In what ways were academic practices of research and assessment reshaped when they were framed through township questions, realities, and commitments, as well as scholarly debate? What was reoriented in urban theorizing when Civic activists and community struggles were recalibrated as places of valid knowledge making? The partnership allowed me to engage such questions, and in doing so, to reflect critically on how my partners and I, together with students, produced knowledge, with whom, and for what agendas.

High Stakes, High Hopes contributes to an archive of alternative kinds of urban knowledges, experiments that work to inspire more varied forms of urban theorizing. Its stories, my turn to narrative writing, comprise ordinary words that become conceptual and can travel. These conceptual tools offer a way to rethink practices of collaboration, teaching, and writing in our field. Urban theorizing in partnership offers ways for urbanists to engage the city, its substance, its stories, its everyday contradictions and possibilities in located and embodied ways. It offers forms of practice, grounded in teaching, to train the next generation of urbanists to understand and engage the city and its urban futures. Through it, I argue, we might in small and incremental ways reimagine the university, its mission, and its mode, embedded in multiple publics and politics across the city.

The Partnership's Cast of Characters

Gerty Square and I were the partnership's core protagonists. Together, we directed the partnership, deciding on its focus, building its work and strategy, its methods and its pedagogies and ways to share our research findings. We created this process step by step, overcoming obstacles, living with irresolvable

6   Chapter One

tensions, and celebrating our successes over the years. We did not have a master plan, an ideal vision, or a timescale. This was work that developed incrementally. It was close to our hearts, often hectic, nearly always deeply satisfying.

Gerty was a strong and courageous woman, who, among many other things, led the United Front Civic Organisation in Valhalla Park. She brought to our partnership a lifelong commitment to justice and to building a fairer, more accessible city. I am an urbanist; trained in geography in the United States, I held a teaching position at the University of Cape Town (UCT). I brought to this partnership my own research interests and questions and a passion for and commitment to collaboration.

Incrementally, we created a way to work together. I oversaw the university side; Gerty was in charge of the neighborhood work. She coordinated our Civic partners, mostly women, a few men, who lived in this neighborhood and worked with her as activists and Civic workers. Fearsome and loving mothers, sisters, friends, "struggle plumbers," land invaders, churchgoers, devout Christians, Moslems, a Hajji, minstrels: these complex, concrete, lived identities shaped their passion for the neighborhood.

As our partnership progressed, we found our roles and filled them. Mina, Gerty's daughter, became our human resources consultant, keeping track of hours worked. Dan was the timekeeper, helping us stay on track in every session. Zaaida was our cook for collective events, the caterer. Koekie loved the camera; when she was happy, we all smiled. Fatima was a constant—quiet and present. Suki embraced this work; it was, she told me, the only time she felt smart, intelligent. Rosemary and Naomi worked with us some years and not others. Daughters joined their mothers. Shireen came in and out of projects, at the start a youngster, by the time we were done, a mom of two, growing up with us across this decade. Our partners' roles were manifold: teachers, translators, guides; they kept us safe from the real possibility of violence, legitimated our presence, tutored us in local protocol, welcomed us as friends, and taught us what they knew, what they loved, tolerated, and hated.

I brought to the partnership my students, at first undergraduates, then postgraduates, who signed up to study the city, to learn urban geography. There were high stakes in this venture, bringing students out of the classroom onto the street, into homes. From varied places and backgrounds, privileged and poor, white and black, foreign and South African, the students brought to this work a richness, wearing multiple caps, with assumptions and feelings, energy, anxieties, and interest. This mix infused our work in Valhalla Park, our readings in class, our writing in journals and papers, the ways in which we shared and explored. Some undergraduate students loved this work and stayed on for postgraduate studies. Some even came back to help: Saskia, Siân, Robyn, Raksha, and Jo.

I started these projects and this partnership a young, fresh, enthusiastic lecturer. My senior colleague made clear to me that my approach to this practical teaching, in the weekly afternoon laboratory session, was over the top. I was "making a mountain out of a molehill" and should not expect extra credit for it, or expect my "colleagues to compensate for these choices." This wise advice referenced the inordinate amount of time that the partnership consumed, what it demanded, the choices I made in this collaborative mode. In the partnership I found a voice, a way to experiment and engage, to root myself in the city and in South Africa, to teach and research. It brought my worlds together, a way to be a white academic, in this inequitable city, in a privileged university, in this period a decade after the end of apartheid. In the partnership work, I juggled my roles as academic, writer, researcher, and teacher, strict and rigorous, quirky and quick, with a sense of humor most days.

The Book's Design

Narrated in stories of the partnership, the book is not conventional. The narratives build a thick, richly layered, and textured understanding of our practice that demonstrates and shows rather than explains. While this book and project have been shaped by scholarly bodies of work that experiment with ways "to do" urban studies "otherwise," these literatures are not at the forefront of my analysis. They do not dictate the book's form. Instead, an engagement with these literatures is boxed alongside the narrative, spaced across the book. Discussion in the boxes places my work in relation to an archive of "doing urban studies otherwise." It shows the work with which I am in conversation and helps direct the ways in which I am thickening the archive through this partnership and its practice. The boxed-off literature discussions offer further reading; they engage the literature; they are part, rather than a disavowal, of the scholarly work I draw on. They situate this literature in the background, threading it carefully into the fabric, the texture, the work of the book.

The book's visuals work in relation to the narratives as well. Like the narratives, they are rich and varied. Working in partnership produced a diversity of intentional and incidental artefacts: from designed books and research products to letters and notes, to syllabi, to events, and relationships. These visual artefacts live in different ways—sometimes in a file in a partner's house, as a map I've kept for another decade on the top shelf of my office bookcase, as treasured letters, artefacts that remind us of our work together. The book

8   Chapter One

shares some of the visual elements of the archive of the partnership. They work as well to visualize and bring into the book those more ephemeral moments, the relationships, the feel, the engagement, the encounters. From photographs, to maps, and so on, they make these moments concrete and intimate. They make visible the materialization of partnership across its practices, across the decade.

To the reader, some of the visuals are directly legible; like some narratives, they offer a precise message and measure. Others offer texture, rather than precise interpretation. In sum, the visuals offer the feel, the presence, the mix that shapes the partnership's work. Like the book's stories, the visuals and their textures are designed to "wash over you."

Each chapter in the body of the book concludes with a coda. The codas are analytical. They articulate crisply "what is at stake" for that chapter. They work as the link to the next chapter. They conceptualize the tools I developed through the partnership, across the book. Lightly threaded, they bring together, rather than foreclose, the argument. In doing so, they embody the complexity and the multidirectionality of the partnership and its practice. Through the codas I step out of the particularities of this partnership, distill some of its learning, and share what is at stake.

In this mix of narrative techniques, I share the partnership's creativity and experimentation, its layered and multiple practices. In doing so the book's form offers an alternative means of interpretation. Through it, I share the journey that brought Gerty and me, the Civic and the university, to work together to form the partnership. The book goes on to track the partnership, its evolution, and its development in the everyday struggles we researched, the city inequalities that shaped the Civic's activism and our research. And, of course, we bumped into contradictions in our work, in the neighborhood and university, in our city. These crisscrossing compromises and complicities shaped our work.

Teaching was of the utmost importance to our partnership. In the book, I trace the ways it oriented us to work back and forth between the neighborhood and the university. I share the research we produced in publications from each project, what sustained our partnership beyond projects and semester-long courses, across the years. I draw the book to a close by pulling together the conversations, the tools, and the concepts that the partnership offers as a means to theorize the city otherwise.

In this spirit, I offer this book, its stories and analysis, as a meditation on collaboration, an ethnography of our partnership and its practice. I offer this book as a change of form, a different kind of academic writing. I offer this book as an experiment, part of an archive of alternative practices that work to reshape teaching and research in urban studies. I offer the book as a source of conceptual tools and collaborative practices for urban theorizing in partnership. I offer this book as a celebration of our partnership across a decade.

CHAPTER 2

A City, a Community, a University, a Partnership

Staying, Not Running

On a damp and chilly Cape Town winter's evening, in an informal settlement community forum meeting, in a school classroom across the road from the settlement in Khayelitsha, the chairman of the forum challenged me with a question. He pointedly asked, "Will you run away like all the university-based researchers we've met before you?" I start this book with this moment, where a research home truth hit me hard.

The chairman's question had two parts. I had come to him and the committee to ask permission to do research, as part of my doctoral work, on his settlement's housing struggles. His question challenged me to account for my research, to explain how my questions were relevant to the settlement's struggles. But his question was as much an invocation "to not run away," to account for what and how I would write, and for whom, to ensure that, at a minimum, the forum would have access to my findings.

Underlying both these questions lay a deeper provocation. The chairman rightly insisted that I focus my research on his housing conditions as a problematic of housing justice. He demanded that I be explicit about the relationship of my analysis to the settlement and its residents' hard daily struggle, to frame my research beyond the metrics of the university and scholarly debate. His challenge questioned the ways in which I would locate my work in the settlement's fight with the city over its legality, in the area's struggle for secure land, in the dreams of settlement families to build a decent place to live, in their aspiration for formal homes on the edge of the City of Cape Town.

This provocation was not unique. Questions like it shape research and its practice in South African cities—and in divided and unequal contexts elsewhere, contexts in which researchers engage and work with activists and or-

A South African Imperative

Urban Studies "Otherwise"

In this work I build on a tradition of anti- and post-apartheid scholarship that, rooted in engagement, aims to challenge injustice and to invigorate scholarship. Immersed in the city, engaged with ordinary people, we might build "meaningful dialogue" (Nyamnjoh 2012, 146) that entangles epistemologies across divides that fracture and interlink city spaces. Ordinary people bear the brunt of inequality and injustice, and, as Ari Sitas suggests, their organizing and everyday life therefore have profound theoretical relevance (2004, 23).

Edgar Pieterse frames the contemporary challenge pointedly, arguing that urban South African scholarship "demands contamination; it demands immersion into profoundly fraught and contested spaces of power and control" (2014, 23), the racialized spaces and processes of the South African and southern city. In the post-1994 democratic period, researchers in and outside of universities have worked actively to engage institutions in and outside of the state, projects rooted in a mix of activist and applied forms of research to create change (see Oldfield, Parnell, and Mabin 2004; Oldfield 2015).

While contentious and challenging, this context pushes the academy to reflect critically on questions of knowledge and its politics. What is, for instance, a socially engaged university (Favish and McMillan 2009) or relevant research and teaching in the contemporary context (Mabin 1984; Lalu 2012; Parnell 2007)? In what ways do scientific practices claim the authority and expertise to "know for," or claim through research to "service" communities, those disenfranchised and impoverished in the past and the present (Oldfield 2008a)? How might we disrupt the invisibility of working in "their" name (Selmeczi 2014), in the name of those with whom we collaborate and engage?

Although not exclusive to South Africa, these questions are urgent in this conjuncture where protest is rampant, democracy nascent and contested, and durable and painful histories of inequality shape everyday life in violent and creative ways.

---

dinary people, movements, nongovernmental organizations (NGOs), and the state. In South Africa the university, its modes of research, its mission and purpose as a public research institution, have been placed under scrutiny. Universities are accused of upholding and reinforcing the status quo, of perpetuating harsh inequalities and forms of injustice that painfully fracture our society. Highlighted powerfully by post-2015 student mobilization for decolonialization, such demands have challenged the university to rethink its epistemological foundations and researchers to reimagine and build public ways of learning and researching. In part, this is a challenge "to think ourselves not apart from the world, but rather deeply and irrevocably caught up in all its contradictory entanglements" (Pieterse 2014, 23).

These layered provocations have shaped my thinking and approach to urban research and its practice. I struggled with the limited and unsatisfactory

FIGURE 1. Valhalla Park, an introduction

Chapter Two

ways my and others' conventional academic practice responded to the chairman's question, "Will I run away?" His question prompted and pushed me. It inspired me to experiment with collaborative modes of research and learning, to build in and on a debate on community-based work and collaborative practice.

A Partnership

Built incrementally over a decade, my partnership with the Valhalla Park United Front Civic Organisation (hereafter the Civic), a township neighborhood-based organization in Cape Town, became a long-term commitment to develop a research and teaching practice in which I could "stay and not run."

A series of collaborative experiments led to my research partnership with the Civic. For instance, at the Urban Futures Conference in Johannesburg in 2000, shortly after I finished my doctorate in urban geography, I listened to an NGO director scathingly ask why universities failed to engage creatively to teach with NGOs working in township spaces and with township organizations. A new academic, a novice in my job, I approached her, cautiously, to ask if we could discuss this idea further. She agreed. For three years my undergraduate students and I worked with her NGO and affiliated community organization in New Crossroads, Cape Town, a township formerly segregated racially as African under apartheid rule with a celebrated history of resisting that regime. This experiment was a first attempt to build a collaborative approach to my teaching.

I explored, as well, ways in which collaboration might anchor my research. In 2000, with a Norwegian colleague, Kristian Stokke, I began a project on the Western Cape Anti-Eviction Campaign and the politics of access to services in increasingly privatized Cape Town. We approached campaign leaders to discuss the research. Without their approval, activists were unwilling to discuss their thinking and their strategies. They were suspicious of academic motives and intent, and of my and my partner's politics. To address their demand that the research prove useful to their struggles, we produced articles based on our research, as well as a report tracking the campaign, participating neighborhood organizations, and their struggles. It was through this research project that I met Gerty Square and Washiela Arendse, leaders of the Valhalla Park United Front Civic Organisation. Initially I interviewed them about the genesis of the Civic, an amalgam of a tenants' association and a concerned citizens group formed in the 1980s in this then-new "coloured" neighborhood produced to the specifications of segregation and the laws of apartheid Group Areas.

My colleague and I struggled in our conversations to reconcile what we

A City, a Community, a University   **15**

did together in this research with what we "owed" the anti-eviction campaign: how we engaged the campaign, what we as researchers had the right or savvy to assert. Awkward, and unresolved in this project, these discomforts shaped my impetus to try something different; in this case, a closer collaboration with the campaign itself. The campaign's strategies were multiple. I worked with activists to develop the Community Research Group, which ran in 2004 and 2005. It aimed to fuse the needs and resonances of activist struggles with the rigor and legitimation of university-based research approaches to support activists' search for information, for data, for ways to articulate their experiences and draw evidence together. This approach to research aimed to support the everyday work that activists undertook to challenge the city precisely and powerfully, as well as to figure out what to do in their areas to solve problems, to strategize around city and national policies, to work together as a movement, no small challenge for activists with few resources.

The Community Research Group included the Valhalla Park United Front Civic Organisation. Through the group, we developed a survey to document informal housing conditions in backyards in Valhalla Park. The research project aimed to bring into the city's view the hardships and severity of informal housing that residents built in the backyard spaces of public housing units. These "backyards" in their neighborhood, fifteen kilometers from the city's center, were "off the map" of the city's housing development priorities; hidden from view, on the one hand, low on the list of urgent housing needs, on the other.

The Civic began an extensive survey of backyards in the neighborhood to demonstrate this housing crisis to the city government. After the first week of surveying backyard conditions, we met in Gerty's small lounge in her home to debrief. Three people sat on the sofa, Gerty in the single chair by the telephone stand, another person on the narrow wooden stairs that steeply ascended to the second floor. I was perched on the arm of a chair, the room in view. "This is so boring," Koekie stated emphatically, when I asked what was wrong, why everybody looked so depressed. Gerty explained, "Look, we've done fifty houses and it took so long. How will we get through the neighborhood's 1,700 households?" I felt the frustration, the impossibility of the survey scale. It was time intensive and tedious. On campus, in the coming semester, I was teaching an urban geography class, for fifty third-year students. I tentatively asked if my students could help complete the survey with the Civic. The others expressed interest.

This was the kernel from which our research and teaching partnership grew. An annual research project anchored the mode of our collaboration.

16   Chapter Two

Three principles shaped our approach. First, the research agenda was decided together. In each project we built from an issue that the Civic was working on, something that social science research could do, that my students and neighborhood partners could do together, a topic with some type of linked urban studies and geography research literature. Second, neighborhood residents (activists, community workers, Civic members), my students, and I conducted research together, a pedagogy through which we learned research skills and the neighborhood and city. Third, we produced publications of varied genres: from posters to popular books, from academic articles to local directories and maps, to this book. Publications aimed to work and resonate in and across neighborhood and university contexts.

The research projects documented and analyzed struggles to access and self-build housing, a critical issue in the neighborhood. The Civic had mobilized to occupy land and build an informal settlement, responding to the severe lack of homes, the realities of families living in overcrowded rental public housing and in shacks in backyards of homes and settlements. The research also focused on struggles to make ends meet, in a context of extreme unemployment and disappearing steady work, real everyday struggles to put food on the table. We completed research on the Civic itself, the wide span of its work, its leadership, its strategies to build a better and fairer neighborhood and city, in the context of varied forms of structural everyday violence and dispossession.

The partnership became a space through which, over a decade, we worked together. The process initiated a different sort of conversation on research, one that was slow, collaborative, ongoing, and which the Civic and I charted together. In this collaborative work—in the annual projects—neighborhood residents and Civic members and students honed research skills and built knowledge. The Civic and I shared our expertise in teaching my students together. The partnership introduced provocative conversations on everyday urbanism, realities that shaped the demands of activism, the hardships of structural inequality, debates on the right to the city, the substance of much urban theorizing.

The partnership built across shifting political and economic contexts. There were moments when development was imminent and services improved. Other moments emerged when a housing project fought for and promised remained elusive, unbuilt, postponed. The university context also shifted. Initially a small project built as part of my own "laboratory," practical work, the partnership work was integrated fully into a curriculum, with its own logic and intent; at one point funded, at another not at all. In incremental steps, across a decade, we worked together, feeling out, sometimes stumbling over, our differences, our commitments, and our aspirations.

Through the partnership, I found ways to refigure the relationships at the heart of my research practice, repositioning my research and teaching. In its always-varied rhythms, in its productively compromised nature, collaborative work offered ways to refigure my teaching and research, to build a foundation for epistemological and political critique.

In this book, I share the evolution of my collaborative practice, inspired by activists and activism across Cape Town. My partnership with the Civic gives the book its substantive problem space (Scott 2014, 157). As a problem space, the partnership emerged as a context and process in which we could work together, in which, as co-leaders of the partnership, Gerty and I could share the

Refiguring

Collaboration and Its Inspirations

In an edited collection on transnational feminist practice (2010) Amanda Swarr and Richa Nagar challenge researchers to rethink the relationships between processes and products of collaboration; to be more conscious of the interweaving of theories with collaborative methodologies; to reimagine reciprocity in collaboration to produce knowledge that can travel beyond the borders of the academy. One way to do this is to "turn our theoretical goals from a 'northern' (university) academic project to the struggles of those with whom we collaborate" (Nagar 2002, 184), practices that Nagar (2014; 2019; Sangtin Writers with 2006) develops in her varied experiments with collaboration. I draw on collaborative approaches to urban research (see Oldfield and Patel 2016), which embrace a variety of forms of knowledge and forms of accountability (Bunge 2011; Nagar 2012).

Yet collaborative work, as a commitment and a practice, is "always productively compromised" (Pratt, 2012). As Pratt demonstrates, theorizing through collaboration is "open to other geographies and histories. It puts the world together differently, erasing some lines on our taken-for-granted maps and bringing other borders into view" (xxxiv). Pratt suggests "what makes such encounters ethically and politically promising, in Sara Ahmed's view, is the possibility they offer to get 'closer to others in order to occupy or inhabit the distance between us.'" Inhabiting this distance, she argues, "opens each person in such encounters to other unheard or unfinished histories and geographies, to other encounters, or in Ahmed's phrasing, 'other others'" (xxxiv). This frame offers a language for collaboration, not as a recipe for useful or equitable work but as always "productively compromised," as a "problematic rather than a social fact" (Chari and Donner 2010, 76), a practice rather than a solution (Routledge and Derickson 2015).

Chapter Two

questions and issues that mattered, the hopes and worries with which we were preoccupied. It was a space in which, with neighborhood partners, we could figure out the means and ways to develop projects, the processes to teach and engage students in varied conversations in different physical spaces—in homes, front rooms, and backyard spaces, in the neighborhood informal settlements, and on streets across the area.

Through the partnership and its processes across the years, we saw and felt the durable legacies of colonial and apartheid rule, physically and racially evident in injustices and inequalities, inscribed in physical infrastructure, in the everyday realities of who walked, who drove, who ate, who was hungry, who worked, who barely made ends meet. In it, we encountered the inspiration and anger that has driven activism past and present, that persists as a search for justice, a legitimate call to arms, a way to reclaim dignity in the city. Through the partnership, we engaged the university, a contested and dynamic site, a product of a racist past, a site where contemporary privileges were easily remade, a site of change, the home of the next generation of students, diverse and dynamic. In this partnership, we came together, we inhabited—and sometimes reworked—the differences between us. On the one hand, we sought, and sometimes found, a politics of hope; on the other, we hit intransigence, the limits of change, the deepening of difference and inequality.

In this spirit, this book offers our partnership as a collaborative problem space. Through it, we might imagine any range of other partnerships, in cities elsewhere.

A Turn to Narrative

I found it challenging to write about the partnership. I grappled with my normal, scholarly, ethnographic, and argumentative mode, its limits, the ways in which it foreclosed the things that really mattered in the partnership. I wanted to write about the labor of love that sustained the relationship, those things that built trust, the practices that enabled our research and teaching to grow.

At a feminist writing practices workshop in Cape Town in 2013, my colleagues encouraged me to experiment with writing stories of the partnership, its incremental building, our movement between neighborhood and university. They introduced me to the narrative turn in the discipline of international relations, to Naeem Inayatullah's approach to the autobiographical, the "I" in "IR," as Inayatullah's pithily introductory edited book is titled (2011).

I found Dauphinee's work resonant and inspiring. A Bosnian interlocuter,

A City, a Community, a University 19

her friend, asked of her, "What do you know about Bosnia? Why did you come, and what did you think you would find?" (2010, 801). In her response to his questioning of what she knew of war, violence, guilt—the focus of her research and career—she recounted, "What expert am I? This is what Stojan Sokolović demanded of me and to which I had (and have) no good answer." (2010, 802) To engage his questions, to respond to his critique, she shredded the conventional academic book on which she was working. Through a turn to narrative, she wrote a different book, *The Politics of Exile* (2013a), because

> narrative approaches allow us to think about the worlds we encounter differently. They allow us to encounter worlds that we normally do not see. They give us different languages and different angles of vision ... Human communication is enframed by these elements that rarely make their way into the texts of our professional lives. And yet, it is this very social world within which our texts seek to produce meaning. (2013b, 348)

Like Inayatullah, I was intrigued by this book and Dauphinee's writing. Inayatullah explains that Dauphinee returned to Bosnia, where "she uncovers an arrogance in her assumptions. She wonders if her research is a kind of violence perpetrated against the fullness of life there. There and everywhere" (2013, 337). He argues that in writing this second book, *The Politics of Exile*, Dauphinee "moves to a form that transcends the usual academic venture but that readers cannot reject as the ideographic portraits of 'mere' fiction" (339). For Inayatullah, "through her interaction with Stojan [the Bosnian friend], the professor generates an intimate awareness. And, by writing through these characters, Dauphinee builds a bridge to herself via the world at large. She constructs a systematic wholeness, an intimate systematicity" (343). He understands her change of form as "a protest against the homogeneity of forms in our field" (339).

Inayatullah proposes that we change form "so readers feel, think, and experience the overlap fictively" (2013, 342). He suggests we fuse strategies and writing from fiction to shift ways of building arguments and theory. Conventional academic modes of writing aim to "make the argument more forceful, cleverer, more anticipatory of reader defenses, or more packed with evidence," more paranoid. Yet this conventional mode of writing fails. As Dauphinee explains herself, "I could not find an academic language to say the things I wanted to say" (2010, 813). The beauty and care, the generosity both of Dauphinee's writing and of Inayatullah's questioning grabbed me.

Dauphinee's struggle resonated. I abandoned my original book proposal, a conventional ethnography, framed by activist-academic collaboration. It did

Chapter Two

not work. Bifurcated, categorical notions and languages of activism and the academy paralyzed me. I experimented with narratives. I found this form of writing my partnership energizing. In this writing I could share the heart of the beast, the partnership itself, what had made the partnership flow, sustain itself; what and how we had learned and navigated, what we had managed to overcome; what we left behind, or worked around. Through narrative I was able to write the layered, nurtured, contingent ways collaborative work and theorizing traveled and unraveled back and forth across the city. The turf of theory emerged, enmeshed in my writing about our work together.

As I was ensconced on sabbatical, caught up in my writing, the solid-state drive of my sleek MacBook Air failed. It had not been soldered sufficiently. I lost all my text, literally and forever. Foolishly, I had failed to back up a word. I slowly recuperated from the shock, recovered from the loss. I wrote this book "again." I signed up for a writing workshop, tentatively offering some initial re-

Experimenting

Writing Urban Studies "Otherwise"

The book's form breaks from the mould of urban ethnography that dominates urban studies. It offers an approach to scholarly writing built in narratives, inspired by Dauphinee's and Inayatullah's autobiographical turn and developed in conversation with Nagar's (2014, 2019), Selmeczi's (2012; 2014; Choi et al. 2020), and Salo's (2004; 2018) work. In this book I draw on stories and on a narrative structure. Stories "'induce feeling,' they 'woo, engage, surprise, persuade, rattle, disarm, or disquiet the reader,'" they generate a particular mood" (Lorimer and Parr 2014, 543). They offer "an acute awareness of the tones and textures, memories and feelings, logics and poetics—of people, places, and times as well as the seemingly mundane truths of life that remain distant or insignificant in the imagination of mainstream academia" (Nagar 2019, 33). They matter, as planning theorist Leonie Sandercock proclaimed: "The way we narrate the city becomes constitutive of urban reality, affecting the choices we make, the ways we then might act" (2003, 12).

As a mode of experimentation, stories unsettle mainstream scholarly forms of writing in urban studies. "The multiplicity of narratives are important for presenting contingent and, sometimes, more uncanny links between seemingly disconnected sites of imagination and experience" (Dickens and Edensor 2021, 17). The work of narrative in this book "embrace[s] 'play and experimentation,' juxtaposition and compression in ways that academic text (normally conceived) does not" (Cresswell 2021, 37). In theorizing through narrative, as Laurel Richardson persuasively suggests, writing is "a method of inquiry, a way of finding out about yourself and your topic. Although we usually think about writing as a mode of 'telling' about the social world, writing is not just a mopping-up activity at the end of a research project. Writing is also a way of 'knowing'—a method of discovery and analysis" (Richardson 2020, 923).

written stories for comment. Through this workshop, in its collegial love and care, I started the book-writing again. We worked together. I wrote more boldly, more bravely. The words poured out, the work of the partnership, its complexity, its richness. Its dynamics unraveled on my screen; the keyboard clattered. I wrote the practice of the partnership in detail, from the mundane to the bold, from the success to the failure. I tracked the partnership, my sense of its ups and downs, the trajectories of our engagements, from our interviews and public events to the field notes and meeting notes. I paid attention to our affect, the evolution, the logics that layered our conversations. In this narrative form, our practice became more transparent, the decision making, its contingencies, the objects and the subject of the teaching and the research. In this writing, what made the partnership tick—our varied roles, our ways of working—came into view, making visible the layered ways in which we worked together.

I drew on my partnership archive to write these stories: the documents, my field notes, research interview transcripts, group reports and project maps, papers, drafts, the varied project publications, collected notes and reflections, and many layers of interviews and conversations with my partners, with Gerty, with students, across and beyond the decade of the partnership work itself. As I pieced them together, they built on memories, on personal notes and diaries, on research letters and photographs, on the minutiae of keeping in touch, on discussions of materials in the process of drafting. I wrote the textured, intimate nuances of this partnership, its forms of collaboration, its smells, sounds, contradictions, and conflicts. Narratives examined openings and closings, contingencies and surprises, places where we felt, saw, moments in which we knew the city, the neighborhood, and the university in banal and remarkable ways, in ways that opened and enriched, in ways that also closed. This transparency of purpose, this fuller accounting, revealed the complexity of choice, of context, of thinking, of doing. As we moved between the university and the neighborhood, building urban theorizing in partnership, we shifted binary notions of who researches whom, of the subject of the research. This shifted the presumptions of expertise, of knower and known in research. In this change of form, I wrote the partnership, our means to weave a way to work together across this city, its segregation and inequality, its varied and deep forms of violence.

Making Visible the Partnership and Its Practices

My change of writing form was not a simple, aesthetic choice. In the narration of the partnership, stories work to make its nuance and texture visible. Storytelling in narrative form brings into view that which constitutes the partnership, its practice, built in and between the categories of neighborhood and university, of partner, student, activist, and resident. It renders this terrain entangled and intermeshed. The stories are a figurative and argumentative way of moving back and forth, through this partnership, across this city. They track the evolution of the partnership in the projects we worked on, the incremental building of our pedagogy, and the partnership itself. They situate the project work and the partnership in the contradictions and inequalities that interlinked and shaped the neighborhood and university, in and across city. They examine the teaching practice across these projects, its planning, its reworking of the syllabus, and the learning it produced. They reflect on the research, its forms and uses in the varied publications we produced.

The stories of collaboration introduce surprises, humor, the openings and closings, the possibilities and limits of the partnership. Stories invite the reader into the partnership, its context, its intricacies, its costs, complicities, and inspirations. They work to make visible a mixture of urban acts and objects. In varied ways they unsettle and reposition arguments about the neighborhood and the university, the complexities of township life and research on it, the contentions of the partnership, its process and politics. Across the book, the layers of narrative intertwine evidence and story, competing truths and conflicting notions, multiple commitments, and forms of accounting.

In writing the partnership, I draw on varied notions of "I" and "we." Stories include me, my partners, and my students. There are moments across the book and in narratives where the "we" is the partnership working as a unit, doing interviews, hosting neighborhood research events. There are others where the "we" is me and my partners, assessing and evaluating students and their research work. There are moments where the "we" of the partnership comprises the partners themselves, negotiating the project in the neighborhood. Across the book I also draw on a repeated and shifting sense of "I," myself. I am the author of this book, visible, a subject of the stories. "These stories interpret and theorize the narrator as much as the narrator theorizes the content and cast of the stories" (Nagar 2019, 33). My curiosities, my interactions, my blunders are woven into the narratives. By design, the writing works to render the "we" and the "I" transparent, through the varied labor, care, authority, and expertise on which the partnership built.

The writing works to reflect the complexities of our collaboration, its practices, our roles as collaborators, laden categories and identities that shift in context, in relation to each other. Collaborators can be friends, comrades, and partners, as well as a vehicle for debate and argument; a context, a façade for broader conflicts. In a more negative and pejorative sense, collaborators can also be traitors, double agents, spies, a difficult, shifting, and contentious terrain. I played many of these roles, as did my partners and students. This mix proved more difficult to write about. It was sometimes harder to see. It shaped how we responded to invitations and navigated refusals. Behind the scenes, this territory shaped what we did, where and how we traveled, together and apart.

The narratives aim to take the reader to a different place, another context, another way of understanding, rooted and fused in the partnership and its collaboration. As such, they summon and enroll; they encourage the reader to think about moments and spaces, conflicts and inspirations. They interweave threads of generosity and trust, heartbreak and love, conflict and compromise, the epistemologies of care and struggle that sustained and invigorated us. They show and share the ways in which, through the partnership, we were implicated in this work together. It is through this narrative approach that I can turn an experiential lens on the partnership, its place in the university and neighborhood, its teaching and research, the foundations from which we theorize the city. In always productively compromised ways, through it we inhabit and rework our differences, we make visible the partnership's flesh, the warp and weft of its fabric.

The way we worked together was no longer method; it became substance. The sharing of the work no longer occurred at the university alone: it was in the neighborhood. The teaching was no longer my work alone, it was joint; my partners taught the neighborhood, they interviewed neighbors. We made mistakes, we blundered, and we fixed. The partnership was substantive and epistemological. Its thick form of implication intertwined evidence and story, competing truths and conflicting notions, multiple commitments and means to account.

Ordinary Words, Urban Worlds

In the narratives of this book are ordinary words, an urban vocabulary incrementally built in the partnership. Ordinary words emerge in the stories through which I narrate the partnership. They arose in conversation, in interviews in homes, in interactions on neighborhood streets, in events in the neighborhood hall; they surfaced in the classroom, in visits and presentations on campus, in the sharing of work, in seminars and assessments at the university. In the stories in which they emerge, they show the genealogies of our partnership practice and its process.

I track the ways they arose in the partnership practice, how we taught and

24 Chapter Two

An Invitation

Urban Theorizing "Otherwise"

In this book I build conceptual tools for urban theorizing in partnership. Through the narratives of the partnership, in ordinary words, I build a mode of theory building (Bhan 2019, 2), "socio-politically rooted in the margins of the everyday" (Dasgupta and Wahby 2021, 420). This work is part of a global call for new thinking on urban futures resonant with the global south as the epicentre of urbanism (Bhan, Srinivas, and Watson 2018), a call for "southern theory" (Connell 2007; Comaroff and Comaroff 2012).

The imperative to engage the city in its complexities and dynamics is an invitation to experiment, to build alternative methods and epistemologies (Roy 2020, 20). It is a means of theorizing that "see[s] value and political power in smaller, 'in- between,' uncomfortable, and multiple theoretical contributions" (Saville 2021, 100). It reflects the partnership's anchor in "insurgent planning" (Miraftab 2009), in partners taking charge, "by dodging, resisting, defying, commandeering, diverting, building homes, earning incomes and attempting in many other banal and spectacular ways to improve their lives" (Ballard 2015, 220).

Building urban theorizing in partnership is a "way of doing theory differently, of working inside out, of fugitive moves and emergent practices" (Katz 2017, 599). In partnership, "what is at stake is not whether but how one theorizes. Thinking of theory in this way affects how we read; the concepts we use; how, for and with whom we do our research; and how, for and with whom we write" (Oswin and Pratt 2021, 595). Through urban theorizing in partnership, I contribute to the call to build alternative cultures of theorizing (Robinson and Roy 2016, 183–186) rooted in "the everyday struggles of urban dwellers" (Peake 2016, 225).

---

learned together, our research methods, each project and its evolution, the publics with whom publications were shared. I map the ways they surfaced in the neighborhood, articulations of (in)justice, the substance of struggles for recognition, an artefact of the city's contested and shifting power dynamics. Rooted in the practice of the partnership, rather than theoretical provocations alone, they are a product of the partnership, its juxtaposition of neighborhood and university, an outcome of our method, its layered substance, its entanglements in practice. Thickly theorized, built in partnership, ordinary words offer a vocabulary, a mode of theorizing (Bhan 2019). They are a means "to generate and imagine theories and practices of the urban" (Dasgupta and Wahby 2021, 420). They are a product of practice, rooted and located, epistemological and political conceptual tools. They offer a vocabulary, a mode of theory building.

Across the book's narratives, I draw from the partnership's ordinary words. They expand the turf of urban theorizing. In flashes of understanding, moments of conflict and compromise, in the hopes that kept the partnership working together over a decade, the partnership offers in its methodology and epistemology a way to create a located, embodied mode of urban theorizing. It is in this spirit and sensibility that I share its projects and evolution in the chapter that follows.

Coda—A Partnership

Initially, our collaboration was a pragmatic solution to our parallel needs. Yet, beyond this start, so much more was at stake. For Gerty and the Civic it was to have an arsenal of research branded by the university to support the issues of their activism. As Gerty explained, despite all their activism and hard work, they still faced a stigma, as a group from a violent, poverty-stricken, gangsterism-ridden neighborhood. The partnership affirmed our partners as residents, community workers, and activists, as researchers and experts key to the making of this neighborhood, to the building of a better city.

Through the partnership, I could build a radical form of pedagogy, one in which I rooted my teaching and research in concrete struggles that addressed and challenged the injustices and the inequities that shaped the city. At stake was a means to make my teaching and research relevant and rigorous, to "engage the city." At stake was an imperative to do urban studies differently by building a form of urban theorizing in partnership, to respond to the deeper, challenging layers that underlay the chairman's provocation: would I stay or run?

CHAPTER 3

A Decade

A Chronology of Projects

"We Need Help"

In Gerty's lounge that cold winter afternoon in 2004, I listened to the report on the backyard survey, noting both the frustration with the scope of the surveying and its companion, a barely touched neighborhood plot map intended to record completed interviews. The survey was designed to count the numbers of backyard structures, map their locations, track the densities of families and individuals residing in shacks and formal public housing units, and record their levels of access to basic services, such as running water and toilets. But the task was huge. There were more than seventeen hundred formal households to interview. Gerty stressed, "We really need help."

I was, then, about six weeks from starting to teach an urban geography class, a course with fifty students, which ideally included some fieldwork. For the class, and for my teaching, working on the survey would be ideal. As the meeting proceeded, I asked, "Could I bring my students on board? Could we work together to complete the survey?" The mood in the room shifted markedly. We started to imagine what such a project might require. This was the point at which our partnership began. Although we did not know it then, this was the foundation on which our research would build for the decade to come.

In this chapter, I introduce the Civic and its leaders and share the chronology of research we completed over the decade, and through these projects, describe the partnership's incremental building. I track the ways in which this initial work grew, the specific motivations that shaped our research, and the ways in which the Civic and I navigated our individual and joint agendas for documentation and research.

Doing Urban Studies "Otherwise"

Colin McFarlane (2011, 2) argues "if we are interested in justice, then we cannot simply ask what specialist and expertise knowledge is and what it does, nor simply how learning takes place—we need alongside this to ask constantly who we learn from and with; that is, we need to attend to where critical urban knowledge comes from and how it is learnt." This is work participatory and community-based approaches take forward "to reframe the production of knowledge by placing engagement and experience at the core of social inquiry" (Shannon et al. 2020, 1153). Cautious of "false distinctions," the notion of the academy as separate from or beyond society, or intellectuals as distinct from social movements, this work roots relevance in "scholar activism" and "public geographies" (Fuller 2008; Mrs. Kinpaisby 2008; Mitchell 2008; Kindon and Elwood 2009; Autonomous Geographies Collective 2010).

Here I build on South African engaged urban scholarship (see Winkler 2013; Bénit-Gbaffou et al. 2019) to reflect on the partnership and its participatory practice. I bring into view the meetings, the teaching sessions, the conversations on the side, the research and interviews, the public and private events, the parties, and research publications. I reflect on writing for the partnership, from the minutiae of memos, the span of syllabi, the guidelines for research sessions, to reports, papers, articles, and books. "The intimate, emotional and embodied relationships, responsibilities, challenges and achievements, that demand attention to more-than-rational ways of being and knowing; beyond the text, the page, the research proposal or final report" (Askins 2018, 1288). I write the practice of "staying, not running," its rhythm and practice, its contradictions and compromises, its politics of the possible, what Mason (2021, 2) suggests is "more than" the research itself. Drawing on his language, "'staying' embodies time and space, built on a sense of care."

The Civic, an Introduction

I first met Gerty Square and Washiela Arendse in 2001 at a Western Cape Anti-Eviction Campaign meeting. Both were community leaders in the Valhalla Park United Front Civic Organisation (the Civic), which at the time was taking part in the campaign. Washiela Arendse was the anti-eviction campaign's treasurer. Forcibly removed from District Six, she and her family were part of the last group of residents moved in the 1980s to Valhalla Park, a then–newly built neighborhood, segregated for families classified "coloured" under apartheid. Responding to the harshness of this forced removal and the hardness of life in this new township, Washiela helped form a neighborhood tenants' association. While serving the campaign's meeting participants from big pots of curry and *akne*, she introduced me to Gerty Square, at the time secretary of the Civic.

FIGURE 2. Civic leaders, partners

30   Chapter Three

A savvy and experienced activist, Gerty was the backbone of the Civic, a powerful figure in the neighborhood. To explain her activism, she recounted her own hard experience of multiple evictions in the early 1980s, invoking clearly and loudly that she was evicted "not once, not twice, but thrice." She described the horror and outrage of coming home at the end of a long day's work in a factory, in Salt River close to the city center, to find her children out on the road, her furniture and possessions unceremoniously removed from her house, dumped on the street. Twice she moved herself back inside her house. The third time she went in frustration to the "rent lady," the neighborhood housing official—an employee of the municipality—to say, "I am a single mother, I earn this much, look at my income slip, these are my costs. I cannot pay the rent; you cannot evict me. Treat me with some respect!" Her personal experience of the hardship and stress of forced eviction was a kernel on which she built her activism, the mode through which she learned to fight and through which she inspired others in her neighborhood. Made redundant from factory work in the early 1990s, Gerty found her full-time vocation in her activism, her Civic work, and in later years, her work on minstrels (known locally as Klopse, or "Coons," a racist term). Over the years of the partnership, Gerty shared her personal story with me numerous times. It framed her explanation of Civic work, a master narrative of sorts, told well, forged in personal hardships, the lived realities central to the Civic's activism.

Through Gerty, I met George Rosenberg, then-chair of the Civic. A stalwart figure in the neighborhood fight for justice, he represented the African National Congress in the democratic transitional local government structure in Cape Town in 1994 and was elected to represent the ANC in the first fully democratic city elections in 1996. Later declared too radical, he was kicked out of both the party and his local government position. A record keeper, he carefully stored and preserved the Civic's history, tracking the neighborhood struggles in files and a photo album.

All three leaders were beacons in Valhalla Park neighborhood struggles. Although over the years the Civic fluctuated in its size and focus, its agenda stayed consistent: it fought for the residents of Valhalla Park. When we first met during the anti-eviction campaign, Civic activists were steadfastly resisting evictions, putting families back into homes, mobilizing fiercely for resident rights, for city resources and support. As a former "coloured" group area, Valhalla Park had formal public housing, and thus its material need for new homes was only semivisible. In consequence, it did not feature in the city's list of housing developments or number among its political priorities. Civic activism on housing aimed to make this imperative visible.

The Civic won some extraordinary struggles against the state, fought in the streets of the neighborhood, in the city's rent office, in the city council in the center of Cape Town, even in the regional High Court. A product of close to thirty years of activism and community work in Valhalla Park, organizationally it was the amalgam of two tenants' committees, both established in the early 1980s to protect renters in this new segregated public housing scheme, and a concerned citizens group, formed in the mid-1980s to mobilize for better conditions, for electricity, and for some respect at the height of apartheid racist rule. It aimed to defend the neighborhood and to fight for better conditions, to challenge political and social engineering central to apartheid segregation, to defend families classified racially by the apartheid state as "coloured," some affected by the last stage of forced removals from District Six in the early 1980s, and others from shanty towns and squats targeted for removal on the edges of the city.

Built in the early 1980s as a dormitory township for families segregated as so-called coloured, the neighborhood was dominated by public housing, attached small single-story three-room homes, and rows of maisonettes, attached two-story units, built with cement block exteriors. In the intervening years, with no new housing opportunities, residents had expanded and extended, building informal structures called backyard shacks by the city, Wendy houses or bungalows by residents. When I first met Gerty and her colleagues, the Civic and homeless residents were in the middle of a legal and physical struggle with the city over their building of a land occupation, a settlement of shack dwellers that had reclaimed a disused neighborhood park.

The Civic had the capacity and interest to welcome the research work I proposed. Gerty appreciated the purpose and potential of research, its possible relationship to her activism, and other possible benefits that might emerge through our collaboration and partnership. Immersed in research on community organizing and urban politics, I was passionate about and committed to teaching fieldwork (something I found missing in my own training). I was excited by and convinced of the potential of fusing my research and teaching interests and had experimented for three years with this form of work with a nongovernmental organization in New Crossroads, a formerly racially segregated African neighborhood adjacent to Nyanga in Cape Town. Through this work, I had developed an initial method and confidently saw the possibilities of embracing this form of collaborative research in my curriculum. I was also excited about the potential of working directly with members of the Civic, themselves residents and organizers, who bore the brunt of the urban political issues that we aimed to research.

FIGURE 3. Between the Civic and the city

Housing Tenure Security 8ste Laan

'It's the only language they understand'

12 July 2005
CAPE ARGUS

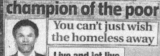

'champion' fights on
Valhalla Park 'Squatters'
Housing the homeless
the RIGHT TO ADEQUATE HOUSING
starts with YOU
champion of the poor
You can't just wish the homeless away
Live and let live.

PLAKKER STRYD

The struggle continues

★ CAPE TIMES THURSDAY, MAY 12, 2005

34   Chapter Three

In terms of our first project on backyard housing, Gerty's and my first objective was clear: to complete the survey started by the Civic.

Getting Started, a Project on Backyarders

Backyard dwellings were technically informal and usually illegal, self-built, or self-bought housing units located in the back and front yards of the public housing flats that most families in the neighborhood rent from the city. These housing conditions were taken for granted in many ways. Backyards were strategic in the apartheid era because they were invisible. By necessity, they were behind homes and walls, built in the interstitial spaces between flats, in the yards behind each one- or two-story unit. They proved a way to extend housing for a family, a way to secure better-located homes in the city despite authoritarian apartheid regulation of movement and residence. Backyards were a form of housing that made it possible for people to stay in a relatively well-located neighborhood near family and friends, near the networks that made life possible, near familiar parts of the city.

In the post-apartheid period, in contrast, the relative invisibility of housing in backyards backfired, so to speak. Neighborhoods like Valhalla Park with their many backyarders fell off the priority areas of extreme housing need because, from the streetscape, the neighborhoods' rows and rows of apartheid-era public housing looked "formal." But enter a home, cross a small living area, head out the back door: it was in these interstitial spaces between the rows of public housing that backyards were largely built. Sometimes they were prefabricated Wendy houses, at others, pieced-together shacks, in which a gathering of salvaged materials made up the walls and zinc sheets weighed down the roof.

Our research goal was therefore threefold: to collect survey data on backyard densities and conditions, to interview families in backyards about their experiences, and to use the research as an opportunity for our partnership to experiment with fieldwork for the first time. We started with the survey, a list of questions that the Civic and I had developed for the Community Research Group work to map and record the geography and demographics of backyards: who lived where, in what densities, and how they accessed basic services such as toilets, running water, and electricity. I approached our then-cartographer in my department, who generated a base map for the neighborhood, which included the city's demarcation of individual plots and streets.

We used the map to divide up the survey area into sections and to allocate a Civic partner and two students to conduct research in each section. In practice, even before beginning the interviewing and surveying, we found the map hard to follow, with few street names included and no landmarks, the normal features through which our partners described and navigated their neighborhood. We worked together to "ground truth" the map, to develop it into a usable guide for the research, and to figure out where area boundaries lay to ensure that each partner was clear on where we would work when the research began.

A logistical plan was also essential. Gerty drew together a team of neighborhood participants, made up initially of those who started the survey work themselves but supplemented by others, largely women Civic members, interested in working on the project. On campus, I organized buses, drew on our departmental fieldwork budget to fund the project costs (in particular the per hour honoraria for the work completed by the partners and the transport costs), and integrated the research into the practical laboratory sessions that I was responsible for in the urban geography course in 2004 and 2005. The work to complete the survey and interviews with backyard dwellers could begin.

Out of the Classroom, into the City

I positioned myself at the front of the Jammie Shuttle (the UCT bus we hired) full of students to direct the driver to Valhalla Park for the first session of our research project. Down the N2 highway that cut across the city, we headed past the Rondebosch golf course, its verdant greens flourishing in the winter rain. We flew past the canalized Black River, a weed- and garbage-choked marker that divided formerly white- and coloured-segregated neighborhoods, in this case, middle-class Sybrand Park from Bokmakierie, full of 1930s public housing stock on the edge of Athlone. As the highway straightened, we continued across the Flats, past Langa, the oldest remaining segregated African area, and Joe Slovo, its dense shack settlement on the highway edge. At Vanguard Drive (renamed Jakes Gerwel Drive in later years, to remember this prominent anti-apartheid activist and post-apartheid government leader), we turned to the left, through Bonteheuwel, another generation of late 1980s public housing, onto Valhalla Park Drive. We followed Angela Street, an artery through the neighborhood, for most students a first view of Valhalla Park, a varied patchwork of fences and walls, front-yard shacks, and small homes. The bus parked on rough part-pavement and road curb, across from Gerty's home. She and her Civic team welcomed us, her yard a neighborhood hub, the shipping container outside a Civic meeting space, with a public phone box built onto the side of her house—in those days, pre–readily accessible and affordable cell phones, a communication line for many families.

Eleven teams of students and Civic members worked together in the months that followed to try to complete the backyard housing survey. We mapped the density, the numbers, and the levels of service access. We discussed with backyard families the lived realities of this form of housing, what was for some a chance to stay with family, for others an expensive private rental, in which they were at the mercy of the formal home's occupants for access to the toilet, to electricity, to water. It was this relational element that made backyard dwelling either a decent option or hellish, an affordable way to stay in place or a costly, worrying form of shelter. Either way, flat-roofed structures invariably leaked, from the roof down, from the ground up, through the walls. Lined with newspaper, roofs held down by rocks and tires, many backyards were little different in physical makeup from shacks in more visible informal settlements across the city.

Our research tracked family histories, rooted in this neighborhood, in the legacies of the apartheid city. Families expected and hoped to secure a stable home after apartheid but faced the hard realities of continued struggle and mobilization, the disappointment of post-apartheid democracy. Our research traced the precarity of living in backyard housing, the everyday realities of limited access to toilets and electricity, to cooking and cleaning. We explored the demands placed on relationships with family and landlords in the front formal houses. In some cases, the research showed the comforts of living with family, of securing a safe space to live in a relatively well-placed location. Many families described their homes as "bungalows," "Wendy houses," rejecting the stigmas of shack dwelling and informality bound up in a language of "backyards."

Many of our Civic team members themselves lived in backyards, a frame that added layers and depth to the conversations in the partnership and with backyard residents we interviewed. For our partner Dan, for instance, this backyard project really mattered. The backyard housing crisis was personal and citywide; it was his story, his own struggle to find a place to live, to house his family of girls decently, securely. He had spent decades of his life in backyards, moving frequently, navigating what he described as destabilizing ups and downs. As he worked with his research team, the interviews he completed resonated with his own understanding and his experience. He shared this thinking, his own hard life lessons, with us.

After the semester was over, we manufactured a large map with backyards marked, tables of backyard data, and a series of reports on "backyard life," which tracked the varied ways in which individuals and families coped with backyard living. We completed close to half of the surveying work and decided to continue in the following year, the next time I offered the class. We learned a lot in this first set of projects on backyard shacks. Completion of the research was much more complicated and time consuming than we expected. Even combining forces, the energy and time of students and the local savvy and knowledge of the Civic, the process required endurance, a system, and a commitment. After this first project, we decided to aim for a simpler product from the research process. Our initial idea was too complex, involving the production of several different types of maps, some for running the project and tracking our collection of data, others digitized and useful—we hoped—as hard evidence for negotiation with the city officials. We rejigged our ideas about what might be doable, useful, and usable.

Struggles for Housing

Housing access is contentious and precarious. There are an estimated 437 pockets of informal settlements across Cape Town (http://ismaps.org.za/desktop.html, accessed November 22, 2021), housing approximately 270 000 people (City of Cape Town 2021). The city's policies toward informal settlements are ambiguous. Sometimes new land occupations are interdicted and evicted, while at other times they are regularized through policy instruments that enable the provision of services—either emergency services as a temporary measure, or sometimes through an informal settlement upgrading process (see Cirolia et al. 2017; Levenson 2021; Ngwenya and Cirolia 2021). Occupying land is characterized by uncertainty: the threat of eviction, the effects of disasters like floods and fire, and the permanent temporariness (Oldfield and Greyling 2015) and difficult living conditions of this form of makeshift accommodation.

Many families opt for backyard dwellings, structures erected within the grounds of existing public housing and spanning a spectrum from rudimentary in nature (what might be called shacks) to formal-looking (the Wendy house or the bungalow). In many cases, backyards are used as an extension of public houses by family members who have outgrown the confines of their family home. In other cases, backyard dwellings are let by house occupants to non–family members for much-needed extra income. Backyarding is unregulated, and so access to formal housing's facilities and services is not guaranteed.

Most residents who live in informal settlements and backyards have registered their housing need with the state (Oldfield and Greyling 2015). The South African Constitutional Bill of Rights specifies the right to adequate housing for all citizens, an important right that reflects one of the state's commitments to post-apartheid redistribution (Huchzermeyer 2001). South Africa thus has an ambitious housing project that funds, develops, and allocates housing to those who qualify according to eligibility criteria, and that provides a basic top structure that the state envisages to be incrementally developed by the household (see Charlton 2009, 2018). Yet, realizing the right to adequate housing is a protracted process and applicants spend long periods of time waiting for a house from the state (see Levenson 2018; Levenson 2022; Millstein 2020).

Chapter Three

We committed to completing the surveying and mapping the following year. Willing to work together again, we had found something productive, albeit slow, defined by the weekly timing of our research and class sessions every Wednesday afternoon, the regularity of the semester, its fixed start and end dates, an ebb and flow with the routine and the pacing of the semester. Our biggest takeaway in year one, perhaps, was how much we enjoyed the process. To mark just that, at the end of the project, we held a braai—a barbeque—at my home, a relaxed moment without an agenda, a chance for me to return the consistent welcome and hospitality, the care I experienced from the Civic partners in this first project. This get-together became part of the project repertoire, a relaxed annual event to celebrate our project's end. In this rhythm, and in the rhythm of the neighborhood, this initial experimentation became a model for our work together.

Land Occupation, a Shift in Research Agenda

In early 2006, I sat in the regional High Court in the center of Cape Town with leaders of the Civic and Sewende Laan settlement families, overwhelmed by the "*milords*"—the archaic language and confusing protocol of the court. As already mentioned, the housing crisis in Valhalla Park had remained relatively invisible to city decision makers and politicians. The neighborhood was not on the city's priority list for formal housing, despite the overcrowding and experiences of homelessness that were common in families across the area. The Civic opted for radical action, mobilizing homeless residents to occupy vacant land, organizing them to build their own homes. Pulling together whatever materials they could find, families had constructed homes in a disused park in the middle of the neighborhood, naming the settlement "Sewende Laan" (Seventh Avenue) after a popular South African soap opera. Civic activism had turned to land occupation. In response, the City of Cape Town had interdicted both the families in the settlement and the Civic, taking them to court for trespassing.

The city's appeal of a judgment in favor of the Civic and families was what we had come to listen to in the court in mid-2006; the city lost its appeal. Sewende Laan families were granted the right to stay in the settlement. The city was instructed to prioritize a housing project for families in this neighborhood and in its surrounding areas.

To reflect this momentous achievement, my class and the Civic teamed up again, this time to document Sewende Laan, the security families found in living in a legal informal settlement, what it meant to families in the settlement

A Decade   **39**

to have fought the City of Cape Town in this epic struggle and to have won, and what this form of Civic mobilization demanded. This land occupation, the building of the informal settlement, our activist partners explained, was a starkly visible way to solve the neighborhood housing crisis, to take the city's mandate to build homes for the homeless into their own hands. Just under a hundred families faced with extreme homelessness came together with the Civic to build homes in 2004 in a dilapidated and disused neighborhood park. As families explained in the project interviews, they could no longer live in the "bush," in the field next to the neighborhood. They could not tolerate moving from floor to sofa, in already overcrowded homes. One family had enough of making do in a rusted and discarded car in a vacant lot nearby.

The city called this act a "land invasion," powerful language that marked this homemaking as illegal. Families described how city law enforcement and anti–land invasion units had responded immediately by calling the South African Police. Together they had torn down these "illegal" structures. The Civic and the settlement families defended the settlement, as Gerty explained, despite "the City shooting at us, rubber bullets; they threw our shacks down with bulldozers . . . We continued building anyway." In response to this persistence, the city interdicted ninety-seven families in the settlement, as well as the Civic, which was named in the court case as the organizing force behind the land occupation. In interviews, settlement families and Civic members described the experience of fighting the city, attending the High Court, working with the Legal Resources Centre (LRC), a nongovernmental legal organization that worked on cases such as this one. In the face of the city's claims that their needs were illegitimate, neighborhood families invited the judge to visit families in their homes in the settlement. They described how surprised they were when he did, the experience of meeting him outside of the court, and its protocol. The LRC, an NGO that tries cases that have potential to substantiate and stretch constitutional rights, represented the Civic and Sewende Laan families. Working with the LRC, the Civic won the case, and the city was instructed that not only would Sewende Laan residents be granted access to services and legality where they lived but also that a housing project would be built in the neighborhood to address the housing emergencies.

This research project was heartfelt, immersed in the settlement families' lives and stories. Each week I joined a research group. In one case, I was ushered in to sit in Aunty Lissie's house. We bent down to enter through its door, were shown proudly around its rooms. We admired the lace, her wedding pictures, and a small vase with a single-stemmed plastic red rose. These personal touches made it her home, precious, in this damp and musty corner of the

FIGURE 4. Mapping Sewende Laan

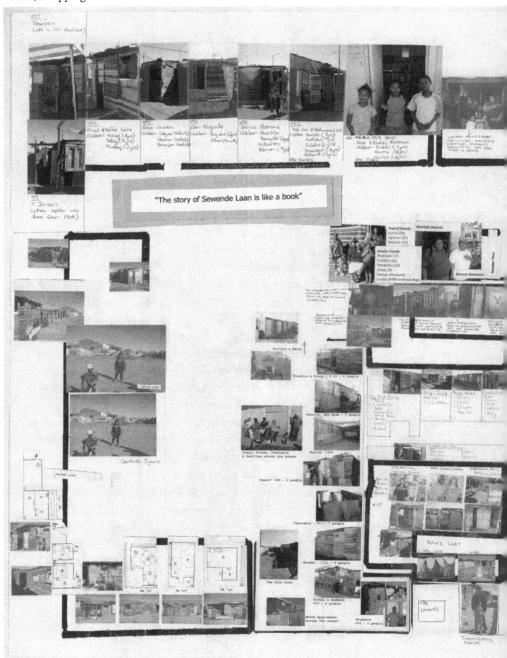

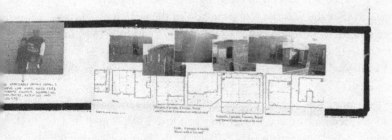
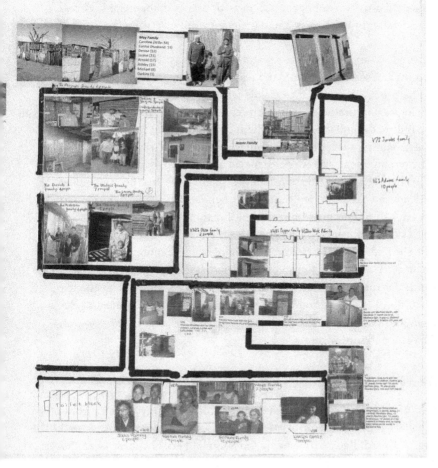

Chapter Three

settlement. We listened and documented her story of housing insecurity, of meeting her husband in the settlement, of building this home, a narrative in which she emphasized that she was the oldest person in the settlement. I listened, helped here and there with questions, which I aimed carefully to build the team's confidence in interviewing.

The Civic and I decided to host a party in the neighborhood to share the work in progress and to reflect upon it, two-thirds of the way through the semester. We asked the students in each group to produce posters to share their work, to show and get feedback on the stories they produced. We hoped that this method would provide an opportunity to account to the families we interviewed. It would also be, we hoped, a chance for others in the neighborhood to better understand the process, the partnership, why we were conducting interviews.

The Sewende Laan research party was a great success. Thereafter, we integrated an annual research party into the research process and the course curriculum. The process helped make visible what the partnership entailed, as well as the logics and motivations of each research project. In practice, these events also indirectly built the energy for and commitment of residents toward civic activism, adding a layer to their mobilizing and organizing.

With the research material, we constructed a map of the settlement, marking each family's occupation, the space they reclaimed. We produced a booklet that shared Sewende Laan stories, the security families found in this illegal act, in the precariousness and possibility of building their own homes. The book shared stories of building and defending this settlement, a complex mix of defending the site and extending beyond the neighborhood, mobilizing with the LRC, traveling to court, the activism required across the years to win the case. The research was rooted in this site, cosmopolitan and constrained; it situated the partnership in these geographies. In the extending of the project to Sewende Laan, our confidence in our partnership method, and in the research and its forms of publication, grew.

Who Has Moved into Agste Laan?

The success of Sewende Laan and the security found in occupying land, ironically, produced a crisis for the Civic when a second land occupation developed in the neighborhood. Backyard residents in Valhalla Park occupied land and initiated a second settlement in January 2006. Families moved onto unused land technically zoned as part of a school adjoining the neighborhood, and no interdiction or evictions occurred. Through the delivery of rudimentary

FIGURE 5. Building Agste Laan

services, the city in essence sanctioned the settlement's permanency. Thereafter families moved into the settlement from surrounding neighborhoods in and beyond Valhalla Park. Named Agste Laan (Eighth Avenue), the settlement comprised families occupying land on two empty fields between the formal edge of Valhalla Park, Modderdam Road (now Robert Sobukwe Road), and Nooitgedacht, the adjacent neighborhood. In this case, however, the Civic had not organized the occupation.

The Civic was anxious about who was living in Agste Laan. How had they related to the neighborhood, to the Civic and its housing mobilizations, and to Valhalla Park families in need of housing? These tensions amplified when the city rapidly provided electricity and essential sanitation services (portable toilets distributed through the settlement). On the new settlement's first birthday, a year after its formation, Gerty suggested we research it: could we find out where people had moved from, their stories? This focus became our 2008 project. As the settlement was unmapped and in the process of being built, we had to revisit and extend our research process and method. The settlement occupied an empty school field and a vacant set of plots next to a city rainfall runoff drainage area on the side of the busy Modderdam Road, a thoroughfare across the Cape Flats to the city's northern suburbs. We had to figure out a substantively and politically acceptable method to engage with this large settlement. We needed to produce a working map, from scratch, a base map from which we could plan and run the research systematically. In a creative and necessarily flexible process, we aimed to produce a sensible and workable base map that could divide neatly into areas for research teams to work on in the coming months. As best we could, we sketched out newly built homes and

44 Chapter Three

streets with a matching narration in words. This penciled and photocopied base map became the structure for our research process, directing research groups in interviewing and documenting work across the Agste Laan settlement. As the map evolved, our logic and confidence in a process for the coming research emerged slowly.

By the end of the project, we had interviewed most of the families in the settlement and produced a full map, which shared the plots, photos of families, our data. It placed Agste Laan on the map, producing a way of knowing the settlement, a collaborative outcome that documented the settlement on the first anniversary of its occupation. The map demarcated homes and recorded families living in the area. The accompanying research shared residents' stories, why they had made the hard choice to move onto the land occupation, what some found as contentious and painful, others as a place of relative peace because it offered some autonomy and privacy. For some new residents, the settlement offered a counter to a past transitory lifestyle that had involved moving between various relatives' and friends' floors in overcrowded public housing. The map marked the partnership, too, its deepening, incremental development, the weaving of its parts together, project by project.

In 2009 we conducted one more project linked to housing, focused on families living in public housing, stories that shared the ways in which families grew across generations, making do in increasingly overcrowded rental homes. In the hardships of these living contexts, multiple generations waited for alternative homes, or, eventually, made a move themselves into backyards or onto land occupations such as Sewende Laan and Agste Laan.

The following year, our research shifted to the Civic and its work across the neighborhood.

Civic Work and Its Wide Parameters

The door cracked open and a head appeared, with a greeting, a request, a question, a need, sometimes a demand. Day in, day out, twenty-four hours a day, seven days a week, the Civic's doors could be knocked on for help and assistance. I too knocked in particular on Gerty's door, with questions on our projects, part of the many layered demands made on the Civic. This often-unaccounted-for work led to a new research project. The organization's work had its exceptional, sometimes spectacular, moments, particularly in its mobilization for Sewende Laan, for land and housing, and its subsequent demand for water in the settlement. Much Civic neighborhood work and activism could also, however, be banal, mundane even. As our partnership developed,

FIGURE 6. What it takes to lead

"Everything was a battle with the council—we had to fight for every one of our rights." -Rosie

"I will not fight a losing battle. We are not losing." -Washiela

"The police can shoot me, but I will never let a person live without a roof." -Washiela

"I enjoy everything about my position as a community worker. I consider it a great privilege and pleasure. It is my core business to make sure that people don't go through what I went through." Gerty

"You are only finished with community work when there's someone else behind you to carry on." -Rosie

"My only preparation to be a community worker was my own experience" -Rosie

"Other communities seek help from the government. We were the first to beat it." -Rosie

"I am demanding. If the Council evicts someone, I put them back in. I don't care what they threaten to do to me because the Constitution says that everyone has the right to housing." -Washiela

46   Chapter Three

we saw this daily—often uncommented-on—work, which demanded consistent commitment and involvement, a permanent, unwavering presence. Neighborhood residents, outsiders, city officials stopped by, asked for help here and there, consulting and checking on neighborhood matters.

We ran a research project that documented and reflected on the multiple mandates on which the Civic worked, the nature of this daily community work and activism. It unfolded in a period in which the Civic worked to define itself. As the organization was caught up in changing neighborhood priorities, its work shifted, as did those involved in the Civic. Uncle George, for instance, sadly took ill with cancer. Eventually, he passed on.

There was a different rhythm to this 2011 project. Each research group chose a theme, something a particular partner worked on, for instance, fighting for homes and preventing evictions. It extended to organizing for access to services and to challenging and circumventing water cutoffs for families deep in arrears to the city. Activism addressed, as well, critical issues in the neighborhood; it attempted to reduce or ameliorate violence, to advocate for better police responsiveness. It engaged with the hard realities of unemployment and the necessity to run businesses, including negotiating informal trading spaces and navigating issues of xenophobia. Other partners were passionate about youth in

Activism as Participation

South African political life and its cities have been shaped by activism. Political activism was critical in the fight against draconian and racist apartheid urban policies and their implementation through segregation and its systemic racism. Many civic organizations and neighborhood activist groups formed during the apartheid era and worked at a neighborhood level as part of what Seekings (2011, 140) describes as the "township revolt" against apartheid.

In the democratic era, these organizations and movements have grown, contesting the narrow ways in which the democratic state has formalized participation processes (Oldfield 2008b, 489; Bénit-Gbaffou 2015). In the post-apartheid context, civic organizations and social movements shift between "invited and invented spaces" of engagement with the state (Miraftab 2004), challenging the injustices and increasing inequality, the state failures of the present era. They occupy the messy, effervescent spaces of encounter with state institutions, holding the state and its democratic promises to account. From boardrooms to backyards, the legal to the illegal, these savvy activities of mobilization, participation, and contestation traverse within, between, and beyond the spaces of the state, the "invited" and "invented" participatory sphere, and in the intimacies of family and neighbourhood life (Oldfield and Stokke 2006).

the neighborhood. They grappled with challenges linked to drugs, alcohol, and teenage pregnancies. Some focused on improving access to health care, and on questions related to and struggles with mental illness and HIV/AIDS across the neighborhood. Others threw themselves into Klopse, into culture and music and sports, and associated citywide competitive networks.

To follow each of these disparate threads of work, we designed an ambitious research process. At times, it felt straightforward. We were well practiced; we knew the routine, from preparatory work to orientation, to running the research sessions, to our process of sharing the work once the research was complete. Yet, in this comfort zone were many little niggles, and frictions, which easily rubbed rough our smooth edges. In its diverse focus areas, the project also opened questions about the Civic itself, its coherence, its identity, the politics of its mandates and leadership.

The research project documented an expansive sense of the Civic's work. In short, it demonstrated what our partners knew: the Civic often faced immense challenges, was immersed in sometimes violent conditions, and experienced limited, if any, state response. The Civic and its multiple mandates were wide, sometimes overwhelming, sometimes dispiriting, often against the odds, sometimes uplifting. The latter aspect was particularly evident in Gerty's and the Civic's commitment to reviving the neighborhood Klopse troupe, the focus of the next annual research project.

Minstrels as Community Development

Nas Abdul Abrahams, the Ward 31 councilor, an elected local politician in Valhalla, a DJ by career, and a long-term resident of Valhalla Park, called minstrels and the Valhalla minstrel troupe "the heartbeat of this brave community." In his analysis, Klopse offered an annual opportunity for celebration, for freedom of a sort, amid struggles to make ends meet. In other words, in his view, behind the public spectacle, minstrels worked to build community. A contested claim, it motivated our research on the Valhalla Park Klopse in 2010, five years after the Civic restarted the troupe. Could our research substantiate that Klopse was a form of community building, of development, activism even? This was the thrust of our research in 2010.

Minstrels are arguably the city's largest single, citywide public expression of working-class coloured identity. A "coloured" rather than "African" tradition, a product of notions of race and blackness bound up in the colonial and apartheid periods, this event is also controversial in the city, easily steeped in racist

FIGURE 7. Competition time for Valhalla Park minstrels

50 Chapter Three

stereotyping. Klopse troupes bring tens of thousands of working-class black people into the city center to watch and compete. The competition is the focus and passion of thousands of residents in over sixty neighborhoods who participate in minstrel troupes, spread largely across the economically impoverished, formerly segregated coloured neighborhoods of Cape Town.

The public citywide celebration of Klopse comprises the annual march through Cape Town's city center on January 2nd, the second celebration of New Year that commemorates the abolition of slavery. It includes as well competing on most weekends in the summer, January through March, in stadia across the Cape Flats. In the project, we researched the work involved in organizing and training the troupe and documented the tradition on which this practice builds, a long and contested history. Klopse embodies and challenges violent practices of slavery, colonialism, and apartheid segregation, and the perpetuation of their legacies in the post-apartheid period.

The research on Klopse proved to be one of our most passionate and successful projects. Alongside the research, my students and I attended competitions, cheering on our partners who performed and competed. One student joined the troupe herself; one year, I participated as a judge. We traveled to Mitchells Plain to the stadium, and joined the neighborhood for fundraising events during the year. Key to the research was documenting the yearlong organizing required to make Klopse happen. This was the fundraising, the training, the choir practices and band practices, the organization of the making of the uniforms, the logistics of transport, the wide array of work that led to a polished passionate performance. Our interviews focused on this community organizing behind the scenes, the daily, weekly, and yearlong work of organizing funding, uniforms, practices, participants, everything it took to run a troupe and compete.

Gerty and Fuzlin, a neighbor and the then-leader of the African Zonks, called organizing Klopse "engineering work." By engineering, they meant the constant effort to hunt down and negotiate bargains in the best places across the city. They laughed as they explained they were experts, trained in making household ends meet and thus skilled at this work as well. Organizing Klopse was expensive and required immense logistics and an events management plan. Our research process documented this work, its motivations, the individual members' work to scrimp and save to be able to pay for transport, hats, uniforms, and incidentals along the way. It explored the leadership behind the organizing, from the troupe's directorship of seven to its eight captains, their organizing members, transport, coaches, access to instruments, and their constant struggle to find contributions (in time, skills, or cash) and to cut costs or cover the excess themselves. Of these fifteen leadership posts, unusually, in

2010, women in the Valhalla troupe filled eleven, with Gerty the chief director (or "owner") of the troupe.

Organizers and their families lived among the paraphernalia of Klopse, the instruments stored in bedrooms, the banners stuck under the stairs, the trophies stacked on the shelf, sequined jackets hanging from windowpanes in small four-roomed homes. Our partner Zaaida explained she dedicated her time, energy, and the greater part of the day "to make Klopse time the most enjoyable time of the year for the residents of Valhalla Park." As Aunty Ellie profoundly suggested when she was interviewed for the project in 2010, this work was "community, what we leave behind, our legacies, to children and great-grandchildren so they can pick up the reins and keep going."

Reflective of the passion and commitment for Klopse, the final research party for this project was particularly full. People we had interviewed were keen to come and engage with the stories they had told us, to celebrate Klopse through the research we had conducted. Used at various meetings and events, the research posters shown at the party were eventually laminated, so they could be better stashed behind Gerty's bedroom wardrobe, ready for the schedule of annual organizing meetings held every year.

Klopse and Its Layers of Politics

The Kaapse Klopse are minstrel troupes that perform in the Cape Carnival on Tweede Nuwejaar, the second of January, a day that commemorates the abolition of slavery. Fused in 150 years of South African history, Carnival includes Malay choirs (Nagtroepers) who march through the city center on New Year's Eve, Klopse troupes from working-class neighborhoods across the city, who march on January 2 in the center of the city, and Christian Christmas Bands that parade in neighborhoods on Christmas Eve. Klopse is the largest single, citywide public expression of working-class "coloured" black identity. Troupes bring tens of thousands of working-class black people into the city center every January 2 and on competition days. It is the focus and passion of thousands of residents across economically impoverished, formerly segregated coloured neighborhoods of Cape Town.

The Klopse tradition reflects and contests the violent practices of slavery, colonialism, and apartheid, particularly the latter's segregation and its forced removals of black people from the city center to the urban periphery (Martin 1999; Bruinders 2006; Bruinders 2010; Bruinders 2017; Miller 2007; Gaulier and Martin 2017). When they march in the city center, what emerges is a "spectacle of liberty and freedom poignantly in the center of the city, the seat and heart of colonial and apartheid power" (Jeppie 1990). Klopse is also a way for residents to express pride in community and neighborhood histories in the contemporary period.

Chapter Three

We published a book on Klopse with beautiful photos and narratives. In the opening section Gerty explained that "when people hear of Valhalla Park, they associated our area with the 28 Gang and drugs. We needed to do something to show the world out there and the people out there—standing in Cape Town, across the Cape Flats, in the surrounding areas—to show them here we are. We come from Valhalla Park. This is what we are doing in Valhalla Park. And this is what we can do." This was why Klopse mattered, why it was central to Civic work: it challenged the stigma she and others felt in the city. In small ways, it helped address the difficulties common in the lives of residents. It linked the Civic and neighborhood families to a proud past. Copies were stacked up at Gerty's front door—the Klopse books were in demand in the neighborhood, in the Klopse Board, and beyond.

Making Ends Meet, Neighborhood Economies

The focus of our final formal research project was a pressing everyday challenge that underscored all the partnership work: the mundane and hard reality that most families in this neighborhood struggled to make ends meet economically, to secure work, and to build long-term livelihoods. In 2012 our research responded to this reality. Two motivations shaped our approach. The first was a chance to document the work undertaken by individuals and families despite the lack of formal employment; the strategies and the challenges to build local businesses, practices that were often subsumed into rough and less helpful notions of informality and illegality. At the same time, the research unfolded in the challenging context of xenophobic violence against so-called foreigners, largely immigrants from other African countries who resettled in South Africa and opened small shops and businesses in many township contexts. In Valhalla Park, as in other townships across Cape Town and South African cities, tensions had fomented; debate on the place of so-called foreign shops and businesses was rife, and sometimes violent.

In this complex and very real context, our project documented and mapped this critical area of the neighborhood economy. We searched for and interviewed the area's businesses, some formal, most informal, some well signposted, public, others less visible behind the doors of homes. Many businesses were run by long-term Valhalla Park residents; others by newcomers, some of whom were "foreigners." We explored how families and businesses started, sustained themselves, and quite often closed in this hard economic context: from tuck shops behind burglar bars, built into a home or an exterior wall, to

A Decade **53**

Struggles to Make Ends Meet

In 2020 the City of Cape Town reported an expanded definition of unemployment, at 29 percent (City of Cape Town, 2020). In neighborhoods like Valhalla Park, the percentage is far higher. With a national economy in recession, and the impact of the Covid-19 pandemic and lockdowns on job losses, unemployment and inequality have deepened (Visagie and Turok 2021), effects felt viscerally by many households. A report from the National Income Dynamics Study asserts that "shack-dwellers faced the biggest jobs slump under the hard lockdown and their recovery has been the most muted" (Visagie and Turok 2021, 2). Despite attempts to integrate corridors of business and trade, Cape Town continues to be characterized by its fragmentation, evident in the stark racialized divide between so-called first and second economies, between the world-class and the ordinary city.

Home-based business and street trading play an important role as a critical space in which residents eke out a living through street trading or small, home-based businesses. Many residents support themselves, working every day to make enough to buy a loaf of bread, to feed their children, to extend credit until payday or social grant payout day. Informal trade is a way of making a living and a way of getting by in tough times (see Dierwechter 2004; Skinner 2008). Yet, neighborhood economies are also politicized. With the increasing numbers of businesses owned and run by so-called foreign nationals, neighborhood economies have provided sites for xenophobic conflict, a harsh politics of insider and outsider, citizen and migrant (see, for instance, Landau 2012; Crush and Chikanda 2015).

---

veggie stalls, open on the street, to seamstresses working on piecework from factories in a home garage, to bakers barely signposted in kitchens here and there, to well-posted burglar bar makers and car repair places. We tracked these practices on street corners, on the main thoroughfare, as well as in the back streets, the garages on corners, the alleyways of the area settlements. We grappled with the geographies of businesses, rooted in the neighborhood but linked to diverse city economies, from sourcing vegetables in Epping Market, the largest city source of fresh vegetables, to trading iron scrap across the Cape Flats, to small piecework contracts with Truworths, a national clothing chain, to one woman's biennial travel to Dubai and China to buy goods for her business.

As in other projects, our method worked with a base map, the foundation of our systematic approach and the designation of research areas for each team of researchers. As always, this strategy proved more complicated in practice, with lots of necessary checking of boundaries, walking the neighborhood, populating our map with useful landmarks so teams did not overlap or step on each other's toes, so to speak. The project traced tricky territory, navigat-

FIGURE 8. Making ends meet

HOUSE SHOPS PARK

Chapter Three

ing both the question of to what degree some businesses were more illicit than others and a heated and dangerous debate on xenophobia, particularly vociferous in that period. From this research, we produced a neighborhood *Yellow Pages*, a small business directory, copies of which were delivered to nearly every household in the neighborhood in 2013.

Incremental Rhythms

Led by Gerty and me, the initial collaboration evolved into a decade-long partnership. The Civic and its community workers, and my class and its students, worked together. Mobilizing against evictions shaped the Civic, its genesis, its leaders, and its mission in this period after apartheid to put Valhalla Park "on the map" as a place in need of housing, first and foremost, and later to rework its reputation as a gang-ridden rough area, where violence was the norm. Our partnership projects followed this trajectory, focusing first on housing struggles, then shifting to the broader array of Civic and community-building work, from minstrels to making ends meet.

There was a necessary slowness to our approach, from our tentative start researching backyard shacks to the completion and sharing of the *Yellow Pages* directory, a publication from our final project on neighborhood business economies. Across the decade, the partnership stretched and expanded. It morphed incrementally and unevenly in the particularities of the projects we worked on, in the comforts and limits of processes, in the often-irreconcilable nature of university demands, in the dynamics of the neighborhood, the ups and downs of the Civic, and in the dissonances of city politics. It was bound up in Gerty's and my commitments to each other, those things that sustained us, that brought us to work together year after year.

The research and teaching became part of the rhythm of activism and neighborhood work for the Civic. Our partnership, our annual class, and my relationships with the Civic and with the neighborhood became part of my rhythm, my research and teaching. The Civic as my partner became a critical interlocutor, key to my thinking and to our collective work theorizing the city together. The annual semester teaching together punctuated the year. This project-by-project incremental approach was essential, developing our capacity to research, to ask questions, to write, and to reflect. In being present, regular, scheduled, in being incrementally and unevenly pieced together, the partnership built and sustained trust and confidence in our process. Civic partners came to present and participate in seminars and course sessions in my department and elsewhere at UCT; they spoke to students, sharing experiences, and our partnership work, with my colleagues. These elements became the foundation of our legitimacy to work together in the neighborhood and in the university.

The process mattered deeply, in its detail, in its productivity, and in its compromises. We designed our projects loosely, building a sensibility of listening, an ethos of engaging respectfully. To participate as partners, we attuned ourselves to be flexible, to listen more carefully, to record and reflect. We paid attention to, and engaged or tried to manage, competing agendas that shaped and split our interests. Students learned to listen in these moments to personal and powerful experiences, to see and to observe, to note and to engage, in ways that illuminated and disrupted literatures and lectures, the traditional staples of university learning. Civic partners introduced students to a world of critical knowledge, what it meant for them and their families, for the Civic, to live and struggle in our city, what it took to engage and challenge these conditions. Partners realized their research abilities. For some this was work that they recognized as always core to their activism. They claimed the title "researcher." For others, research and its pedagogies were new, a different way to engage with their neighborhood and the city, to link to the university; and, sometimes, to re-see themselves.

The partnership held together our varied rhythms, interests, and capacities—contained in the research process. It allowed us to work in and against the grain of neighborhood agendas, the urgency of crises, the imperatives of the Civic, and its struggle with the city, as well as in and against the logics of the university, its prerogatives, and practices. It built on the trust we nurtured to maintain and extend an unconventional research and teaching mode. We innovated and persisted, sometimes deepening our approach. At other times, we just kept going, working in and across the urban inequalities that divided us, living with the conflicts that so easily could have torn us apart.

In many instances, relationships became friendships, which enabled us to work together, to sustain the partnership over the long term. In the fabric of these growing relationships, over the years we celebrated birthdays—my daughter's and Gerty's grandchildren's birthdays, Eid and Christmas, special moments like weddings. Crises also galvanized us, such as the devastation of fires in Sewende Laan and Agste Laan that razed homes and lives repeatedly—shattering losses. In later years, increasing levels of violence refractured the neighborhood. Gerty's son-in-law described an afternoon when the police forced him, together with Gerty's then six-or-so-year-old grandson and others, to lie flat on the street outside her house. He emphasized the humiliation of this act, of being searched in front of their home. These dynamics placed our partners and neighborhood residents on constant high alert, a vigilance

Chapter Three

and violence linked to the police, a citywide underworld, a politics beyond direct neighborhood activism and mobilization. Other moments were hard and intimate, deaths in families, my father and Gerty's father passing on. Attending her father's funeral, I saw her face and her passion, her power and energy, in her son and her brother as they gave the eulogy, in the proud picture of her father, placed on his coffin. She could not attend my father's funeral, held outside South Africa, but she probed the shadows of my sorrow, helping me navigate my sadness.

We planned a final project that would focus on the long-organized-for housing project once it was built. We intended to interview Sewende Laan families following their move out of the settlement into formal homes. This last project would be, we hoped, a chance to celebrate and engage this key outcome, the product of Civic housing activism. We hoped to understand and document its meanings for families, this transition to home ownership, the product of moving into these long-sought-after homes. But the housing project was delayed, once, then again, and again. The bulk infrastructure, the water and sewerage pipes, the roads, were laid. The construction of actual homes— the top structures—was first delayed by city bureaucracies and funding cycles, then caught in a neighborhood-city rivalry, a contestation for construction contracts, a politics in which the project foreman was shot, reappointed, and again shot at twice. The contractor withdrew. The project was postponed, then cancelled, then eventually rebudgeted and scheduled. (Only in late 2021 did the process of building homes finally begin.)

In this increasingly contested, unsafe neighborhood context, the weight and responsibility of our partnership became too much to bear for our partners. The neighborhood's social fabric frayed slowly, at first. Then, at increasingly frequent intervals, families became housebound, shots ringing out at random; uncertainties multiplied. Such heartbreaking hardships and violence exceeded our partnership and any issue our research might have addressed. Our work together slowed down after 2013, eventually ending in 2015. While we remained in contact, long-term friends, our formal partnership and collaboration concluded.

Coda—A Process

To keep the partnership relevant, we had to shape-shift with its politics. We began small, in some senses, ambitiously in others. We incrementally extended and developed the process. We built the partnership, solidified it, worked on it, tinkered at its edges, changed its middles. As the Civic's work shifted, so did the partnership. These changes were substantive, our mode changing with the projects we worked on. Housing questions required a particular approach, a mix of guidelines, a set of experiences. Klopse and Civic work required a different array of links into the neighborhood, a tailored mode of method and writing, a journaling process. Across the slow time of a decade, the Civic's role altered and transformed, its form changed. Across the years, the partnership changed, as did I. Over the long haul, the relationships we built exceeded the partnership itself. Our partnership became a friendship. We developed ways to extend and translate this work, a way to root it, to sustain the partnership across and in the contradictions of the university and city.

CHAPTER 4

Crisscrossing Contradictions, Compromises, and Complicities

Navigating

Contradictions shaped the partnership. To work together, we navigated tensions in the neighborhood, in the university, and in the city. We had to confront or work around things that were hard to discuss, that could not be spoken about. We saw, felt, debated, and engaged crass inequality, deprivation and wealth, comfort and hardship, the violence and inequality that fragmented our city. To work together, we navigated a conflicting and challenging, iniquitous, and sometimes-violent terrain.

Our neighborhood partners, for instance, had to navigate our presence, the logics and purpose of bringing me, and my students, into the neighborhood, into homes and local intimacies, a presence that clashed at times with existing neighborhood hierarchies and demands. Students discovered difficult truths in the neighborhood and navigated the contradictions of the academy. I was thrust into positions of authority—sometimes beyond my expertise—and supposed neutrality, drawn on to be loyal, sometimes outside of my comfort zone. The mode of our partnership sometimes collided with university processes and norms. The Civic, its activism, and our research partnership were located in a conflictive and contested neighborhood terrain, as well as in a broader—sometimes violent—city politics, a terrain and politics that far exceeded the partnership itself.

To sustain the partnership, to complete our work together, to bring about productive outcomes, we had to compromise. But compromises, the choices and complicities they elicited, were risky. They were full of hazards, embedded in contexts that were hierarchical and conflicting, contested in varied ways. This broader terrain exceeded the partnership itself.

Between Refusals and Invitations

Eve Tuck and her coauthors Mistinguette Smith, Allison Guess, Tavia Benjamin, and Brian Jones argue that work between universities and communities "require[s] an ethic of incommensurability" (2014, 57) because "the Academy's colonial history and future ... contours the power imbalance that persist." In contexts of settler colonialism, like South Africa or the United States, "solidarity is an uneasy, reserved, and unsettled matter that neither reconciles present grievances nor forecloses future conflict" (Tuck and Yang, 2012, 3).

Tuck insists, therefore, that collaborative work is always "contingent" (2009, 57). It is wrought with "refusals," "not just a 'no', but a redirection to ideas otherwise unacknowledged or unquestioned" (Tuck and Yang 2014, 239). The process of collaboration is "a series of encounters across our many differences" that "offers proof of the possibility of bringing together and sustaining a relationship with those who do not share an identity, but rather a commitment to work together towards loosely framed, continuously evolving, common ends" (Pratt 2012, xxxiv). In navigating in and between invitations and refusals, collaboration offers possibilities for "accounts for the loss and despair, but also the hope, the visions, the wisdom of lived lives and communities" (Nagar 2019, 417). In collaboration, we must "linger with" refusals (22) to understand what they embody, what messages and intent they might carry.

I draw on this conceptual work to figure ways to track the partnership, its practices of engagement and dialogue, its forms of accounting and consent. In writing the contradictions that shaped the partnership and my experiences of it, I work "to represent structures of violence without reducing them to accessible narratives that re-enact the very violence that 'we' seek to confront" (Nagar 2014, 13–14). These tensions raise profound questions for academic integrity and the forms of accountability and politics that shape research work.

Ek Is Die Baas!

As we took a seat in Gerty's living room, Masnoena rushed in. She pushed her way into the center of the small room. A quick exchange proceeded in Afrikaans; voices rose. Aunt Gerty interjected: "Ek is die Baas," I am the Boss! A tense argument unfolded about Agste Laan and Valhalla Park, about who had the right to work where. Was Agste Laan part of the neighborhood, the Civic's turf, its territory? Shouting ensued. The air could have been cut with a knife. Gerty turned to me and said, "Sophie, please leave. Just leave now. I will sort this out and I will call you. This is nothing to do with you. Just go, now, immediately!" I left, stung by this sharp and radical turn to what had felt like a productive afternoon.

It had begun well. With some visiting colleagues, I had traveled from cam-

Chapter Four

pus to Gerty's house in Valhalla Park, zipping down the highway to Modderdam Road, onto Valhalla Park Drive, entering Angela Street, driving past the library, the clinic and community center, the fish shop, over the speed bump, and through the traffic light to reach Gerty's home. We were meeting to start the preparatory work for our forthcoming project in Agste Laan, the relatively new, and rapidly growing, informal settlement in Valhalla Park.

With jackets and layers on to keep the chill out, as a group, on foot, we headed to Agste Laan. The settlement was nearly a year old. We walked through it, around the edges and in its interstitial spaces where people were building homes. A young man hammered in pallets to construct his wall; another fixed his roof on this cool but sunny winter afternoon. A middle-aged woman was connecting her home to electricity, intertwining and duct-taping a precariously strung cable. Our aim was to map and demarcate areas for research, not an easy, or straightforward, task.

There was no map for the settlement, so we started from scratch, roughly and approximately. Some streets were obvious, well established. These were the places we started. Stone Road ran parallel to the formal roads of Valhalla Park. It was the first place where families built, the site where the settlement started. Long Street, a diagonal dusty road, was demarcated, sufficiently wide for vehicles to make their way across the field. It stretched from the edge of the settlement on Modderdam Road to its far edge, which backed on to the formal houses of Valhalla Park. At the junction where Long Street met Stone Street, a gathering site marked the space for a weekly food pantry for settlement families run by a Valhalla Park mosque. It was another useful landmark on the map. We followed and traced out Lorna Road, a well-established but narrow footpath named after Lorna, a founder settler, whom we met sitting in the sun outside her house. We maneuvered around the tarmac of the former netball court, around which families had neatly built homes. In this manner, we made our way through the settlement, drawing and marking the hard features on our base map, determining sensible boundaries to designate research areas, writing thick descriptive notes to describe them: "By the house with pallets, up to the polka dot fence." Led by Gerty, we spoke with residents as we cut through not-so-clearly demarcated front and back yards, clambering behind and between shacks with our papers, scrawling and tracing out an increasingly detailed map. We explained to residents about the forthcoming project, the students we would bring with us, our reasons for wandering in and out of the settlement.

The map we traced and built would be the template on which the research would unfold. Eventually, we were satisfied that it was workable, reliable

Crisscrossing Contradictions   **63**

enough and navigable by the research teams of students and Valhalla Park partners who would start the following week. We returned to Gerty's house, making our way back along the rough, sandy stretch of Stone Road, moving back on to the harder surfaces of the formal part of the neighborhood. As kids headed home from school, we passed Shruu's Tuck Shop, Aunty Fadielah's shop, and the cheek-by-jowl maisonettes that populated the formal part of the neighborhood.

But, back at Gerty's house, at the end of the afternoon, the process had gone awry. The project had been stopped. Gerty had sent me home. I waited it out, worried and stressed. Practically, on the one hand, my course depended on this project. It was at the core of my curriculum. On the other hand, this conflict felt nasty and difficult, and I cared a lot about the people involved. In the meanwhile, I delayed the project start, informing the students, in part, about what was going on.

The following week I received a call from Gerty. "Soph, can you call the councilor please?" she said. "Nas Abdul Abrahams. He wants to speak to you." I agreed. I asked Gerty if she knew what he wanted to speak to me about. She did not respond directly with details. In the meantime, before I had a chance to call Nas, Masnoena telephoned. She told me in no uncertain terms, "Sophie, look, you cannot come with your students, you cannot. Agste Laan residents will barricade the neighborhood; they will barricade the entrances to Agste Laan. They will. You cannot come. I just wanted you to know this. I want you to know that they will toyi-toyi, march, against you." Oh! I was nonplussed and shocked. "How did this come about?" I asked Masnoena. "What is this fight about? Please help me understand." She too responded opaquely.

That evening I called the councilor. He replied in a friendly tone of voice, "Yes, Sophie, hello." He explained, "Look, my dear, I know that you are doing good work. But really, we did not struggle for democracy for this. We did not. You do not know with whom you are working. They, the Civic, are not a representative community organization. They are not. They claim to be, but that is not the case. So let us be clear, you cannot work in Agste Laan. You really cannot. You cannot continue in the way that you have done in past years. You cannot give the Civic, their friends and families, jobs. This is not how we work here." The concrete issue that had catalyzed this broader politics, in part, revolved around who had the chance to work on these projects. Who had access to the pay, a small but significant sum, compensation for twenty hours of work, perhaps enough money for some groceries to feed a family for a few weeks at most? In these circumstances, with work scarce, the matter of this access had blown up, expanded into a neighborhood issue.

64   Chapter Four

I discussed the problem with the councilor and the possible ways we might go forward with the research, suggesting, "Could I perhaps come and see you in person? Could we not hire some Agste Laan people to join us on the project? Could this be a way forward?" Perhaps, Nas replied, adding he would get back to me later with a response. Gerty and a few of our Civic partners met with the councilor and with leaders of the settlement in the days ahead. Later, they described this meeting as a fight, a storm of accusations. The outcome of the meeting, however, was a clear and productive compromise. Gerty contacted me to present a way forward. Could I add six people to the roster of Civic activists working with us? These six new partners would come directly from Agste Laan.

I could.

Six additional settlement partners were recruited and introduced into the project and our process. Aunty Fadielah was the chairwoman of the Agste Laan Committee, a shop owner and resident in Valhalla Park next to the entrance to the settlement. We were also joined by Andy, a resident of Agste Laan, a flamboyant cross-dresser with immense personality, who carried us all along with his humor and spark throughout the project to come. Sylvia was recruited, a tall woman with presence, one of the founder members of the settlement, her hard life written on her skin. Nawaal joined us, a quiet and shy then–mother of two who lived in the settlement at the intersection of Stone and Long Streets. The last addition was Shereen, a bubbly, rotund woman, who embraced us with her good energy and who brought along her friend and neighbor to round out the group. This officially selected group of Agste Laan residents merged seamlessly, as far as I could see, into the Civic partnership.

What drove this set of negotiations? There was certainly more to this conflict than the politics of hiring partners. A range of legitimacies came into question: a mix of politics and identities, of claims to represent and the right to authority. Gerty felt her legitimacy challenged as head of the Civic, her status as a legitimate leader of the neighborhood placed in question at a moment in which the neighborhood had expanded. Its limits and territorialities had shifted in the process of land occupation and the building of the informal settlement. At stake for Masnoena, for Aunty Fadielah too, was an assertion of leadership, of an attempt to stake a claim to be recognized. Others were organizing less visibly to get access to jobs, to have a small opportunity to link to our project and partnership, to the university perhaps too. But the most likely reason for this organizing was to access the small financial compensation, the pay that this particular job opened up. The councilor, a Democratic Alliance representative, a man formerly of the African National Congress, thrust his oar in to assert his legitimacy as the leader of the neighborhood, as the official voice, the sanctioned representative of democracy, and the man with the neighborhood at heart. Where was I in this mix? A vehicle, a vessel for competing agendas, for these negotiations and debate; I was both present and absent, symbolic perhaps, both contested on the one hand and personally acceptable on the other.

Delayed by a few weeks, the project relaunched with a welcome to the Agste Laan participants and reintroductions in the settlement itself. In the following weeks, I reflected on my ready agreement to this compromise, made under the threat of expulsion from working in the area, from our project going ahead. On the one hand, our new partners from Agste Laan were keen to participate and key to the undertaking of careful and legitimate research in the settlement. They felt like an excellent addition to the team of partners. Yet the politics of this argument and my own positioning in it left me needing to know more. I delicately probed further. Was Masnoena a turncoat of sorts, telephoning me neutrally, yet stoking the fires of discontent in Agste Laan? Was this part of her personal frustration with Gerty's powerful presence, her authority on matters "Civic"? What was Nas's agenda? To reassert himself in a place where he was in the fabric, but not visible, part of the neighborhood, a long-term resident, but in need of a clearer political presence himself? And what was my own position in this mix? How should I understand the trust that I thought we had rebuilt? I was only tangentially part of the conflict, for some a scapegoat, an entry into an argument over what was a long-standing tension and dispute. What were my own interests? On a practical level, I had to make the teaching happen, with class days and the semester rapidly passing. More importantly, I wanted to sustain my research partnership with the Civic and, perhaps more importantly, with Gerty, my close collaborator and friend.

In our productive compromise, I became part of a solution, a vehicle for a debate, part of a broader turf and territorial argument. I did not hold power or authority in this context. My academic authority, its legitimacy, sat somewhere far on the periphery of this conversation. The debate had to be navigated. There was a cost, but it was absorbable, the compromise was productive. In the longer term, it made the partnership more solid. I was told later that the argument cleared the air to a certain degree in the neighborhood, too. It did not address the turf, the legitimacy question, but it made other conversations and working relationships possible.

Don't Worry Lady, I Have a Gun, I'll Shoot!

These were the words a policewoman said to me, as I and the three other judges of the Klopse competition were surrounded and body-guarded off the field in Mitchells Plain, late one unseasonably crisp summer night. Herded carefully, shuffled to our cars, parked strategically (as I realized later) next to the dirt road, an unofficial exit out of the stadium that bypassed the now angry crowds of Klopse troupes, competitors discontent with our adjudication.

This was the end of a six-week adventure of sorts. I had been a judge for the Klopse Board, organizers of the competition in which the Valhalla Park minstrel troupe took part. The board had invited me to judge, despite my concerns about knowing little about Klopse. No, they had assured me, "You are neutral. You are from the University of Cape Town. You will be excellent." I had been allocated the judging of the troupe banners and the boards, constructions along the theme of the carnival that each troupe designed, built, and held aloft on a pole at the front of their march. I had helped too with the assessment of the Klopse Jol, the march that each troupe performed to demonstrate their unity, discipline, and rhythm. The other judges, all of them musicians, assessed the singing and the music, a more detailed and difficult terrain about which I knew nothing.

We had sat each Saturday for a month, under a canopy at the front of the stadium, as the troupes competed. This competition had reached its culmination, its apex, that night, marked by our move from the canopy to a caravan, in the simulated shape of a Castle Lager six-pack of beer. This was where the night had gone wrong. The results were expected imminently. Yet, it turned out that the chief adjudicator had not added up the results weekly. We were "locked" in the six-pack of beer, desperately adding up result numbers, figuring out the order of prizes. Much had gone amiss. I had my results completed, but really the board and the banner were the minor prizes, not heavily contested and relatively easy to assess. The music was a different story. The minutes ticked by; half an hour gone. I sweated. The temperature inside the tin can rose; the crowds outside were increasingly vocal. The master of ceremonies reassured them that surely "the judges will soon be done."

Two long hours later we emerged. It was pitch dark outside; the troupes merged around us; a circle cleared on the field where the Klopse Board organizers stood. They handed us a microphone. I started. For the banners, winner of third place; second place; first place . . . Then the music awards were announced. One troupe kept coming first. Disbelief was visible all around. I watched my friends in the Valhalla Park troupe grow fidgety at first, then outraged. Gerty stood in front, arms outstretched, legs planted wide. Her presence grew, seemingly to contain the anger, to stop her troupe from revolting. I was a target, one of four. A sinner, a traitor, I was tarnished with the same brush. I had betrayed their trust. How could Aziza, the young girl who sang a solo for the Valhalla Park Troupe, not have won, she had a voice like an angel, she was even bringing out a CD? How could, how could, how could? The questions multiplied, they resounded and echoed around the dark stadium. What was I doing here? The tension amplified, time slowed down, and then sped up, as we were surrounded and shuffled off the stadium grounds.

As I drove down the sandy dirt road out of the stadium and then accelerated down Vanguard Road out of Mitchells Plain, I felt an urgent need to get myself home, to insulate myself from the raging critique I knew was stirring in the buses as they rattled down the road back to Valhalla Park and back to neighborhoods, from Delft to Athlone, and across Mitchells Plain. Forty minutes or so later, I arrived home. I felt finished. "How did it go?" my family asked. I recounted the contestation of our adjudicating. My husband, a Capetonian, laughed and then explained that all Klopse results were contested, fought over, always the height of controversy. In other words, this conflagration was par for the course.

I returned to Valhalla Park three days later. It was one of our last research days for our project on Klopse. Our partners were subdued. They were angry with me but kindly trying to pretend otherwise. The questions and the interrogation seeped out in muted form as the afternoon went by. "What were you thinking, Sophie? Were you helping them cook the books, is that what you were doing for two hours?" How could you be complicit in that? That was the bottom line. The neighborhood assessment was that the judges had organized for X Troupe to win overwhelmingly. This was the traitorous act of which I had been accused and convicted.

My protestations and explanations did not amount to much, although everybody assured me that I was right in my judging of the boards and banners (my partners excelled in these categories). I reflected on my assessment, my own predisposition toward Valhalla Park, in this juggling of roles as judge, research partner, and friend. These details were intimate to me, part of the research we undertook in this period. They shaped my thinking, my own loyalties to my friends in the troupe, to their commitment to Klopse, and the hard, almost impossible, organizing work it took to make it happen.

A few weeks later, I was called to the adjudication assessment meeting. It took place on a hot Sunday afternoon in a crèche next to the Joseph Stone Auditorium, on the corner of Jan Smuts and Klipfontein Roads in Athlone, about halfway between Valhalla Park and where I lived. I was worried. I actively dis-

Chapter Four

tanced myself from my musical colleagues, even physically removing myself to the end of the trestle table. I accounted for my judging when called to do so. I needed to salvage my partnership, redeem myself, move forward productively. My narration of events separated out the "banners and the boards," my judging task, a strategy that emphasized subtly and, at the same time, starkly that I was not "with them," the other judges. Conscious of my own maneuvering, I cast myself as innocent in the conflict at hand. I watched too as my friends, the Civic partners, collectively and, it seemed to me, quite systematically, disrupted the meeting. Zaaida asked the critical questions, in a voice of steel, quite unlike how I knew and thought of her. Another partner caused consternation and raised his voice. Gerty stormed out. This was a well-orchestrated and effective dismantling of not just the adjudication but also the board, its legitimacy and reputation as a committee capable of organizing a decent Klopse competition.

It was foolhardy, a mistake, to put myself in the position of "judge" and adjudicator. Why was I surprised that my judgments clashed with our partnership, with our liaison and its commitments, the critical collusions and loyalties that sustained its working? Yet, this politics was not actually about me, and some of it was not about the other judges on the panel, sitting further down the trestle table. It was a politics of community organizing, of Klopse, the Civic's struggle to assert its voice, to maintain legitimacy in the neighborhood and in the city.

To whom was the academic accountable? What roles did I play? Never the objective neutral researcher, in a partnership I was partner, friend, and researcher. A colleague, and every now and again, a judge.

Disquieting Differences in a Wilted, Waterless Garden

The law enforcement vehicle, in tandem with the subcontracted company paid to disconnect the water, had pulled up on the curb outside the township house. They had taken out their equipment, opened the water meter on the street, and publicly inserted a stopper to limit the household's access to water. Neighbors and residents had observed; the family felt humiliation. They could not afford to pay either their water debt or the reconnection fee. In the neighborhood and in the Civic, there was an argument about what such families should do.

Could they survive on the water dripping into the bucket, slowly, all day long? The immediate effect was a water shortage: the inability to do laundry, the need to limit cooking, to cut out cleaning. This change in habits was hard within their home and painfully visible to neighbors. In the longer term, their pride and joy, their garden, wilted. It had been planted and nurtured by the family's deceased grandmother, the original occupant of the house—the family felt torn, hurt by the desecration of their grandmother's memory and publicly humiliated as the garden died slowly, in view, day after day. They lived on the drip, limited to a bucket or two of water a day. They felt the paralysis and disempowerment of such limited access, the private hardship and the public nature of their cutoff and its effects, its emotional consequences.

Should the family reconnect illegally? They explained that they were fearful, conscious of the possibility of legal recourse and criminalization. Nearly everybody had water debts and hardly anybody could reconnect legally by paying off a portion of arrears and the reconnection fee. Some families chose to live on the drip, while supplementing their water access. They lowered their heads by going to the informal settlement next door, a place where there were standpipes and water was not metered. They requested permission. They begged for access to the water tap, carrying water back to a formal home, feeling individually the humiliation of this "step down" from formal service. Others suggested that families should live within the free basic water allocation, the fifty kiloliters per household the city allocated without charge, conserving water usage, individually embodying the city logic of "careful" use, of living "responsibly," within a person's means, as a "good citizen" should. For some Civic members, for Gerty and many of our partners, the solution was obvious: "Reconnect, it is so easy. Know your rights." You just need a "baboon spanner" and a "struggle plumber" to reconnect.

Sitting in Gerty's lounge, we discussed these competing ways forward, this debate in the neighborhood and city. Gerty told us about "the long stories that people tell" and proclaimed that "you shouldn't be ashamed," but she was not in an easy spot herself. She was positioned ambiguously. As a formal representative on the city's ward forum, its subcouncil, she could not break the city's laws. She was caught in a game in which she could not reconnect publicly and so sent residents to others for help. In the subcouncil, she could report maintenance problems and water leaks, but she could not challenge the city's water policies directly. Neither could she challenge the broader social discourse that a person was a criminal if they did not pay, or an irresponsible citizen if they were not "water-wise."

The two students working on the project felt torn by the debate, caught in the moral recriminations thrown in each direction by our neighborhood partners. They were taken aback that Fatima, their liked and respected neighborhood research partner, and Gerty took opposite positions. The students wor-

70   Chapter Four

ried about the research, about the pain the questions evoked, about the debate in response to this family's and others' suffering. They felt caught too in the hard impossibility the family faced living on the drip, the impossibility of their living conditions, making do, by the city's design, with only a dripping faucet for a water supply.

This research project revealed the pain characteristic of the debate in which it swirled. We were complicit in the inequalities and injustices we lived with, in my—and in most of my students'—easy, taken-for-granted access to water; in the things that could be sorted, compromised productively, even if it was difficult; and in the conflicts we had to live with outside our control.

A Partner, a Land Invader, a Ward Forum Member

Uncle Dan was dictating his assessment of his student group to Zaaida. It was only in that moment that I realized he was not confident in writing. I came across them sitting at a coffee table in the departmental tearoom, relaxing before the final student paper session they had come to the university to attend. Their conversation looked intense, a quiet but focused back and forth. Zaaida was the scribe, carefully recording what he wanted to share with his students. He had been a partner and participant since the start, a key cog in our partnership. Year in, year out, I had handed him an assessment sheet. Unintentionally, I had excluded him. In this spare hour, he had the space and time for a little privacy to complete the assessment in conversation with Zaaida. I blushed; I hoped they had not noticed me observing. There were so many things I might have misread, things I realized I did not know.

Dan was a deeply religious man: faith and his Christianity permeated everything he did. He was timekeeper on our projects, a gentle disciplinarian, keeping our groups organized and accounted for in every project; checking our departure and arrival times, keeping us honest about the work we had to complete. Gerty's right-hand man, Dan had been integrated into the Civic's work, drawn in, one of just a few men who were actively engaged on a daily, weekly basis, year after year.

Over many conversations over the years, Dan had shared in his quiet way slivers of his life. His hard struggles living in his mother's house in a backyard, on the one hand a parent and husband, on the other still a child in his mother's home. His subsequent many moves from backyard to backyard, the insecurities that marked him daily, his subservience to those in the house in front, at their mercy for access to the toilet, to water and electricity. He embraced our research on backyard life because this was his story, his experience. He found

Crisscrossing Contradictions   **71**

validation in documenting his and others' experiences: committing to moving into a settlement, to building a home and fighting to stay in what became Sewende Laan, these acts made him an activist. Simultaneously, he was a land invader, the city's tag for his search and claim to have a home of his own, a place where he could rest and raise his family of girls, and his youngest, a son.

At the end of a research session, after the students had piled onto the bus and departed, waved off and out of the neighborhood, Dan told me quietly that he could aspire to be the councilor. He was clear that he did "council" work in his monitoring and maintenance, in the guidance he offered to neighbors, in his leadership roles in the Sewende Laan Committee and in the Civic, and, of course, in the ward forum itself. He was so pleased to represent his community. But he felt a deep-seated tension, particularly in relation to the city's code of conduct. He explained, "I had to sign that I would not go against city council policies as a ward forum representative. But this I can't do. I am representing Sewende Laan as well as Valhalla Park. I am a land invader, and a ward forum member. I have a home because I invaded land. I have a secure place to live with my children and my wife, because I was part of the Sewende Laan struggle. But I am a ward forum member too."

He signed the code of conduct but was not willing or able to revoke or cover up what he was required to do to live, to sustain himself and his family, to build his community. Against "council policy," they were acts that meant he had, in fact, broken the law and the code of conduct. This mix and its irreconcilable tensions troubled him.

A land invader, a ward forum member, a cog in the network that sustained the neighborhood, a shack dweller, a neighborhood partner, a father, a man of the church. These roles sustained and legitimated him. In between were contradictions that shaped what he could and could not do. On this intimate terrain, and through this long-term engagement, I was conscious too of what I could and could not know, of the assumptions that shaped my work in this partnership despite my best intent, of my and the city's complicities, the contradictions we made visible in this work together.

Fear, the Complicities of Xenophobia

In Agste Laan, two Somali traders fled out of the back entrance of the shack, running for cover when we arrived to interview them. The research group, the students particularly, were devastated. Their presence, their wish to interview these traders, had generated fear. Literal and epistemological violence suddenly held visceral meaning. The home-based business research project was

Chapter Four

completed in the context of a violent and jagged debate about the place and legitimacy of so-called foreigners to run small businesses. In the context of massive national tension, xenophobic signs were all around us as we mapped out local businesses and interviewed families about their histories and struggles, what they faced making a living in this informal and small-scale economy. There were no easy divisions between the politics "out there" and the politics of our partnership. No easy smoothing over or reconciliation of this angst and anger, the "us" and "them"; simmering, sometimes under the surface, sometimes violently present. Across the partnership, we held different views on this debate.

Our interviewing was tense, however carefully we tried to craft it. Should so-called foreigners be allowed to work in this neighborhood, to run their businesses? Some of our partners were vehement: foreigners operating businesses in this neighborhood were not legitimate businesspeople; they should not be there. Others appreciated that foreigners needed to trade, to do business, to work; they also valued the cheaper prices that these new informal shops offered. They liked the array of goods they sold.

Everywhere we interviewed, neighborhood businesses were struggling. Social welfare grants were largely the only source of income, a small trickle of pennies and rands patching up household budgets. Households had to make more money, to try and sustain a business and obtain the necessary cash to put a pot on the table. But local, long-term, resident-run house shops were not competing on prices. Their costs were too high. They struggled to stay viable, to stay open. Many had closed. One city narrative was that small "local"—in this case, Valhalla Park—business owners had been outcompeted by newcomers. But they could not afford to stop running their businesses in the neighborhood. And, as the story became more specific, its angst and anger took shape, as did its target: Somali businesses, "foreigners," interlopers who had infiltrated the neighborhood.

This powerful strand of anger is what drove Somali traders to flee their shack when the interview team arrived. A few months after we completed our project, the tensions exploded. Petrol bombs were thrown at five Somali shops, some in rented front rooms of homes, others in shacks in front yards; the shops were burned to the ground, left as shells, with blackened windows and damaged interiors, the businesses eradicated violently. In the aftermath, a Civic partner described these ruins as surreal. Another partner disputed the "petrol bomb" language the media had used to describe the expulsion of Somali traders. Another claimed it was an internal neighborhood fight. It may have looked like xenophobia, but in fact those traders rented from a notorious

Crisscrossing Contradictions — 73

gang leader. And those families that had allowed them to locate a shop in their front room or to put a shack or container on their property, well, they had no choice. They owed this man, this gang lord, money; they had to fund drug habits. These rumors and interpretations, these complicated tentacles, linked tensions together. The gang lord was not someone to mess around with; after the bombings, he was convicted and jailed for four murders in the neighborhood. Was this xenophobia or a turf war, a product of gangs and the competition to sell drugs, internal to the set of networks through which the traders had come into the neighborhood?

The Civic held contradictory views on these issues. They were, at times, the face of resolution, negotiating and brokering peace, and of negotiation. I arrived at Gerty's home one day as three Somali traders emerged with her from her bedroom. They had been negotiating bringing in a new shipping container to sell from, discussing where it might be located, and its hours. In the moments before the petrol bombing, the Civic tried to find some resolution; it warned traders that tensions were increasing. They let foreigners know that it was time to leave, to get out, that "we cannot protect you anymore." Yet, in another moment, the Civic claimed it had founded, even inspired, the xenophobia, and it had done so to protect locals, to address the literally physical and emotional challenges of putting food on the table, of making sure that family businesses did not close, that "foreign interlopers do not steal these opportunities away." Sometimes the same Civic members soothed and stoked these flames—tensions that drove and quelled this violent debate.

Was this project a mistake? Was it a political error to do this research on local businesses, in this period, in this partnership? Were we not complicit in these conflicts? We had a moment of relief, partial success in the neighborhood-based party at the end of the research project. We had invited everybody interviewed to join us. One group had gone out of the way to really express their concern and interest that a Somali trader, whom they had interviewed, attend the party. He was hesitant. He did not agree to come. Then, Angelo, our partner Margie's son, arrived with this man and his wife. They entered late, quietly. Angelo had persuaded them to join us at this neighborhood party in the local crèche, despite its potential unpredictability. He guided them to a table. The students who had invited them brought the family a plate of food and sat with them. It was a moment when it felt like our interviewing had not been so damaging, like it might have built a little thread of something. In this context where nothing was clear, where much was violent, this couple was quietly and visibly present. They were part of our conversation.

Discomforting complicities were intimate and powerful in this debate, in

Chapter Four

its contestation, in our partners' and our different positions. Our project itself exacerbated these tensions; it was shaped through them.

A Complaint

Our teaching and research process required all sorts of forms of management and gatekeeping. Gerty and I were joint keepers of our partnership. It worked in part because our turf and territory were clear. It worked because we trusted and respected each other, because the partnership and its projects were important to both of us.

But, today, I was alone in Gerty's house for the first time ever, entrusted with Wafeeqah, who was then nearly one, on my lap. I was the leftover babysitter, not by design but by necessity. Every adult in this house had to fill in for missing neighborhood partners. We had arrived from campus, each student research group ready to head out to do household interviews. Yet not all our partners were there. A number were missing, including Gerty. She had been at the day hospital and was now in the bus company offices sorting out the buses for Klopse. These were all crucial things, emergencies, everyday needs, urgent community business.

Somebody had to stay home with the baby. Leaticia, her mom, was at work elsewhere. In this moment I was that "somebody."

I sat with Wafeeqah. I had never been in this house by myself when nobody else was home. It felt strange. The normally busy and bustling house was quiet, silent. Sandwiched in a row of upstairs-downstairs maisonettes, the house shared walls with neighbors on both sides. I could hear the neighbors through the kitchen wall, people passing in and out on the street outside. I felt torn and a bit vulnerable. I tried to pretend otherwise: to be relaxed, to play, and to sing a song, to entertain this little girl. On edge, I waited for someone to return to the house so I could hand over the task of babysitting and get back to the research work.

Before we departed back to campus, I raised the query of who had missed the session with Mina. She was our timekeeper, our HR expert, we joked. She kept the records of who had attended, what time they arrived, what they should be paid. By design, she was my first port of call when this sort of issue arose. Mina passed the message on to Gerty, as a complaint.

Gerty requested that I remain behind at the end of the session, after the students returned to campus in the bus. After everybody had departed, she called me into her bedroom, a small slice of personal space in this busy, well-used house. Although I had sorted out payments there and observed others enter-

Crisscrossing Contradictions

ing and emerging for past meetings, this time was the first instance I had been requested to meet there. It was a moment when she called me to account. At issue was my critique, through Mina, of the missing partners the previous week, implicitly, of her own absence.

It was not an easy conversation. Its subtext was a debate about who worked on these projects, who had entry into the partnership. This question was Gerty's turf, located in her role of drawing together a team in the neighborhood. Across projects, and across the years, there were ebbs and flows of partners working on the projects. Koekie had a job looking after her granddaughter once her daughter returned to work. Lefien moved in and out of projects, sometimes busy with piecework seamstering, outsourced from downscaled textile factories in Salt River, Cape Town's old industrial hub. Rosemary, Aunty Washiela, Naomi, Masnoena, early project stalwarts, moved on, in part a reminder of their busy lives, in part a reflection of shifting roles in, and relationships with, the Civic. I shared my concern about Gerty's role in the project, too. She was part of a research team with a student, but did she have time for this role, to work with him? As a rule of thumb, these negotiations and decisions were Gerty's territory as coordinator of the Valhalla Park team. Gerty was clear: I had overstepped. I had trodden on her turf, her right and need to choose who worked, to assess who was best, what her role should be.

We both backed down. We reached a safe, productive, compromise: she could play a different role in this project. One partner was taken off the list; an uncontentious choice as he had to work so could not really participate anyway. Two additional people were asked to join the project, brought onto the list, one to replace the departing partner and another to be a backup if Gerty could not be at a research session herself. This solution we could both manage. These compromises made the projects work. They were also a sign of wear and tear, the labor of running the partnership, the small dynamics that needed to function inside the partnership itself. Not a simple question of logistics, or management, of doing this or that, these dynamics shaped trust, confidence, what we could and could not know, could and could not see, or make visible. These layers, this work, lay at the heart of the partnership itself.

Didn't You Wonder Why?
Neighborhood Crime and Violence

"Didn't you ever wonder why the Civic do not run a neighborhood watch? In Valhalla Park, surely this is an obvious thing to do. There's so much crime." The vice principal of the local high school tossed this pointed question at me

Chapter Four

as we stood together outside the Cape Town Television Offices, where he had taken part in a panel discussion against the planned closure of his school by the Education Department. He insisted on repeating this rhetorical question again: "Haven't *you* wondered why they don't run a neighborhood watch?" Valhalla Park resident Ashraf nodded his head as the vice principal provided the answer: "The gangs won't let them, they can't. The Firm won't let them. They are the real authority in this neighborhood." The Firm was the syndicate of powerful businessmen (and some women) who ran the city drug trade, a network that linked the neighborhood into citywide channels of selling and using.

These tensions subtly arose, when, for instance, students engaged critically with issues of gangs, gangsters, and violence in Valhalla Park. "Who told you that?" was a common response. "That's not right, we—the Civic—resolved the gang problem." As problems with violence increased in the latter years of the partnership, these responses were harder to sustain. The most powerful gangster from Valhalla Park, the head of the Firm, Colin Stanfield, passed away from cancer. Though he stayed in the leafy suburbs in Rondebosch where I lived, his family was resident in Valhalla Park. There he was a hero, a protector, a patron, a source of school fees, a facilitator and funder. In the city beyond, he was a gang boss, a scoundrel to some, a known criminal who had served prison time and avoided other charges, the leader of an illicit network of drug dealing. The Civic was one small local cog in a broader context, one in which the Firm was powerful, operating at scale, across a broad urban and regional territory.

Was this conflict and the Civic's position as simple as the vice principal's account would suggest? The Civic could not operate in opposition to the Firm, but this neighborhood politics was complex and layered. The Civic members worked with Stanfield; they relied on and deeply appreciated his support. To show their respect, and as a mark of his and their centrality in this neighborhood, they organized the transport for his funeral, close to two hundred buses that brought people from all over the city, from the region beyond, to the Valhalla Park sports field where a memorial was held, the site where the housing project would eventually be built.

In the intervening years following Stanfield's death, much changed. His sister held the reins of his empire in the immediate aftermath, but she herself passed on. As control devolved to the next generation, there was a fight, a struggle for leadership, a splitting of this family legacy. For a period, it tore the neighborhood apart. The official Civic narrative asserted gangsterism had been solved in the neighborhood, as the organization itself brokered peace in

1994. They called the competing gangsters to account. They negotiated an accord, one that held up to a point. Yet, this narrative frayed around its edges, depending on whom we spoke with.

The Firm was present across this neighborhood, a publicly unspoken set of links and connections. A more subtle distinction was at work, one that separated out those who controlled the drug trade, the Firm, from those who were gangsters, involved in violence and criminality. This was a topic in our partnership often swept under the rug. The vice principal elaborated, "Look, I know the Civic has been widely successful in housing. I acknowledge that. But, you know, the gangs and the Firm don't give a damn about housing." His message was clear. The Civic was only one of several entities, a small one, operating amid a set of powerful forces, competing powers that shaped and broke the neighborhood, shaped and broke young people's bodies and minds. The vice principal's concern reflected the pandemic of petty crimes and burglaries fueled by *tik* (methamphetamine) addictions in the neighborhood and city. More immediately, he was motivated by the dead body he had found outside the school gate as he had arrived at work the Thursday before we spoke. The spate of murders occurred with increasing regularity, a hard and harsh reality that increasingly shaped the lives of our research partners, the neighborhood, and this part of the city.

An Endpoint

Could we sustain this partnership, this way of working together in this progressively violent context? With increasing shootings, with the fraying of gang leadership, and the proliferation of increasingly younger recruits, we reached a point where we drew our research projects to a close. The pressure and risk had become too great for our partners. The Civic could not bear our weight, the responsibility of navigating us safely through the neighborhood.

Funerals punctuated the weeks and months, the latter years.

What next, who next?

These repetitions were the rhythm of our updates. Innocent bystanders dead. Shot doing normal daily things. A young boy caught in the crossfire; in the shop down from the hall, the night before the High Tea, the Civic fundraiser we attended. Oscar Loggenberg dead. He went out to buy coffee and sugar from a house shop around the corner from his own home. Shot in the leg first. I'm told he called out, "I'm not a gangster," and was then summarily shot in the head, left to die on the road. His wife heard the shouting. She heard the shots. She did not think it was him. His father-in-law found him lying on the road, dead, covered in his own blood, another life taken away. A neighbor saw the shooting, she ran after the gunmen, the boys who shot Oscar. She followed them into Sewende Laan, lost them in the settlement. She knew their names and was willing to talk, despite the risk. She was a brave woman, a neighbor, and a friend. Uncle Charlie and Aunty Doreen—stalwart Klopse and Civic members—Oscar's parents, were heartbroken. Uncle Charlie never recovered. A few months on he passed away, too.

Umar, our partner Fatima's son, was shot in 2014. He went to the shop on the street corner to buy his mother a small birthday present. He was shot dead on his way back home. I called Fatima the afternoon after his funeral. She was flattened, devastated. A few months later, our partner Aunty Fadielah's son, Mogammed, was killed. He had been shot in Shruu's, her shop, in the front part of her house. She had to keep working there, living there. She lost her husband to murder in this house as well.

Old wounds split open, new wounds created.

Heartbroken, Aunty Fadielah had to keep going.

How did you go on when your son had been shot, walking to the shop to buy you a bag of chips for your birthday? When your son had been shot, on his way home from work, walking across the field, caught in the crossfire when looking up to see where gunshots had come from?

Shot, gone, forever.

Families devastated; lives lost, forever.

Zaaida asked me, rhetorically, "How do you walk down the street outside your house when a twelve-year-old walks past you carrying a gun, when he checks you for looking, for noticing? When he talks to you like he's in control, when he shows you his gun again when you try to put him in his place?" How do you go on when a boy in his early teens is shot sixteen times in the head, shot because he belonged to a gang, shot in retaliation for his own violent, heinous, brash acts? How do you go on when a kid's impetuousness, twelve-year-oldness, is held in check, embellished, and destroyed by a gun and by the bullets that shattered his skull? How do we make sense of kids killing kids?

How do we make sense of this violence: at night, in the morning, next door, on the street, on the way to school, to mosque, on the way home from church? Ten or so short kilometers from campus, from my home, a world away.

Eventually, the partnership came to an end because of a world bigger than it, a broader topography, the conflicts that shaped this neighborhood, city, and society.

An Empty Fridge

I hunted for my water bottle before leaving Gerty's house at the end of a research session. It was hot and I needed a quick drink before I headed into the traffic and home. I had left it somewhere. It was not by the sofa or on the display unit by the television. I checked in the kitchen, it was not on the counter, by the sink, or by the stove. Leaticia called out from upstairs: "Soph, I put your water in the refrigerator so it will be nice and cold." I opened the fridge, bare empty shelves. I opened the freezer, where my recycled water bottle sat, a solitary item in an utterly empty freezer; a dire shortage of food a hard reality in this home.

I could not refuse this stark reminder of material realities in this home. I could not refuse the inequality that underlay the partnership, a consistent and critical reminder of the limits of our collaboration and research. However creative, incisive, or productive, our work together could not address the material differences, the economic inequalities that divided us, the inequities of income and life that we navigated in moving between university and community, between my home and this home in Valhalla Park. These were ever-present inequalities, ever-present struggles. These contexts and their hard tensions shaped our partnership. We found and felt them out, we meandered through and stumbled over them. They emerged in the ever-extending and always partial ways in which we came to know each other and work together.

Coda—Contradiction

To keep the partnership going in the neighborhood meant working in and between invitations and refusals. We worked amid real conflicts. A research lens on a question could divulge, lay bare, expose. We sometimes caused harm in exploring questions. It meant working on some topics and not others. We could turn some everyday contradictions into research questions. In some we could see and acknowledge our complicities, the tensions that the partnership generated. Some were conflicts from which we had to look away, too dangerous to touch, too dangerous to research. Yet, whose questions counted? To which invitations did we respond? Whose refusal was reckoned with, when? When did refusals become visible, a conflict, something to engage with, something to avoid?

The contours of our research shifted in the contradictions between the research lens and the contradictions and compromises of politics and activism. In this mix of epistemology and politics some things became clear, others remained opaque, unnamed, not known. In teaching through the partnership, we found ways to work within these tensions.

CHAPTER 5

Teaching and Learning

Across the City, Back and Forth

Onto the Bus

In the privileged comfort of a big blue University of Cape Town Jammie Shuttle bus, we headed down the N2, the campus receding in the distance, the mountain at our back. It was the beginning of the semester, a first class visit to Valhalla Park, the bus palpably full of nerves. Past new and old housing developments, we turned off the highway onto then–Modderdam Road, left onto Valhalla Park Drive, then down Angela Street, into the neighborhood. As we pulled into the fenced parking lot of the public library, we saw a small crowd, our partners, who had gathered to greet us.

At the small neighborhood library, then quite newly built, school kids were busy with homework while an old man read a newspaper at a small table across from the entrance. We funneled through the turnstiles into the meeting room on the side of the public area. Civic members organized the chairs, stacked up around the room's edges, setting them out in circles. We sat, layered, knee-to-knee in this intimate, stuffy space. The Valhalla Park Civic participants introduced themselves. Some quietly nervous, others confident, some clearly curious, the forty or so students then shared where they came from, both afar and close by.

Years into our projects, my partners and I knew the orientation routine. We felt confident in the process. The frame of the course shaped the schedule and rhythm of this work, with projects organized around the thirteen or so weeks of the semester and the once-a-week afternoon block, from two to five, in which the class was scheduled. Most of the participating students were registered for a semester-long, credit-bearing course at UCT, part of a human geography curriculum in a degree in environmental and geographical science. Initially this pedagogy was part of a second-year course called Cities of the

South. As the years progressed, it moved to a third-year course on urban geography, and then became part of my postgraduate teaching for fourth-year and master's-level students. In its earlier forms, at second and third year, it was a mandatory course for students enrolled in my department. Undergraduate courses were team-taught. My component was initially the practical session, allocated to questions of method and fieldwork. Traditionally, long practical sessions were designed as laboratories. Each Science Faculty class was allocated a three-hour afternoon practicum scheduled for student lab work. This is the slot I repurposed for the partnership work. Later, I claimed a block in the course, as a way to build lectures and seminar discussion around the neighborhood work directly. In its postgraduate iteration, as a full semester-long course, I experimented most fully with mixing the research and scholarly reading with learning and teaching in collaboration with the Civic.

The curriculum included weekly journal writing and two assignments: the production of a poster and an academic paper. The posters were shared at a research party in Valhalla Park three-quarters of the way through the semester, an annual event in our partnership process. We started this tradition in the Sewende Laan research project when we invited everybody interviewed in the settlement to share the research with us, to check and celebrate their stories of pieced-together, hard-fought-for security. Research parties were bustling, positive affairs. There was something hectic, warm, urgent, and communal in this celebration of our joint work. The posters were a way of sharing a first layer of our research, our initial analysis and our interim findings. They drew on field notes and interview materials, on the conversations that brought each team together, on the layers of local interpreting and knowledge that situated the findings. They were a portal of sorts, a way of sharing our research findings, a way of building the arguments later made in publications. They included maps and photographs, a first take on the stories that each research team formulated from the detail of interviewing, observing, and interpreting neighborhood dynamics.

Following the research party, students worked on their academic papers, which they presented in a formal set of presentations in the final course session on campus. These papers were assessed by myself and our partners, as well as—in earlier years—my university-based colleagues who co-taught the undergraduate courses with me. The routine of this schedule and the imperative that it function, that the class run, offered a productive structure that shaped the process: the orientation and its prior preparatory work in the neighborhood and on campus, the intensive six weeks of research in the neighborhood, the writing and reflecting built into student work and paper development, and

FIGURE 9. Teams of researchers

Chapter Five

lastly, the sharing, assessing, and disseminating of the research at the end of each semester and project.

Intertwined in the process, in shared expertise and knowledge, our pedagogy was our compass. Our collaboration helped us to question what we knew. It challenged us. It extended university notions of critique and truth. It brought the city and ordinary people into the classroom and made their homes, streets, the neighborhood a teaching space. In building relationships, we could be adept and informed theoretically, steeped in everyday struggles, engaged in rigorous fieldwork, able to confidently build and reassemble ideas about the urban, to move back and forth between these poles. We steered literally and conceptually between the everyday and the idea of, and the necessity for, reading the field. Through this pedagogy, students were introduced to different kinds of knowledge and expertise. They became immersed in the realities and struggles of the neighborhood, realities welded with an urban geography literature and theory. This mix gave rise to new, syncretic forms of knowing and meaning making, immersed in urban praxis, in this city, fractured in its inequalities, its fraught and inspiring everyday realities. In this way we moved from learning theory to theorizing.

Teaching was the compass that set our bearings, made us oriented to and able to navigate this neighborhood, this life, this family, this home with its leaks and difficulties, this Civic struggle, these particularities. It located us: we learned and engaged the city through Valhalla Park, through intimate household struggles and inspirations. We experimented and innovated, while holding in productive tension the university and neighborhood and their competing and challenging demands and needs. The pedagogy was hard and embracing. It was rigorous and compassionate. It splintered and shattered stereotypes. It was teaching as joy and inspiration. It brought relationships and identities to the forefront and prodded and pushed, massaged their engagement. The pedagogy was political, but not preachy or didactic. Its politics were crafted and found, sought in the trajectories that the partnership brought to the fore. Felt, lived, experienced, and questioned, the partnership's politics were found and nurtured in moments of conversation and observation, in engagement with each other.

A Gangster Snap, a Zoo

Conspicuous was the word. One student in particular caused consternation. She had not read the syllabus or attended the introductory lecture; she did not

FIGURE 10. Running and recording fieldwork

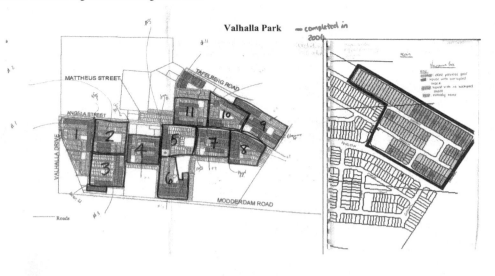

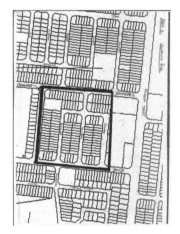
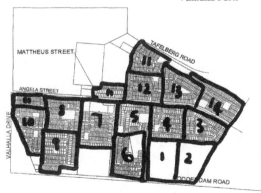
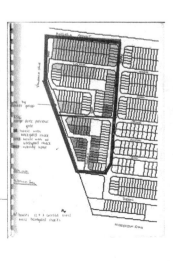

86   Chapter Five

know we were going "out" for our afternoon practical session. It was our orientation session, the first visit to Valhalla Park for this group of third-year urban geography students. The plan was to meet in the community library, share introductions by the Civic on their activism, the area's history, then walk around the neighborhood, to take that first look and to team up students and neighborhood partners in small groups, the teams that would conduct the research together for the rest of the semester.

In the preceding weeks, I had met with the Civic and those working on the project to discuss and refine the focus and our method, a strategy for the research, and to think forward through any issues that might shape or disrupt it. These conversations were logistical but, critically, also substantive. They were a process in which we reflected on what the project might mean, why it mattered, the history of the issue in the area. These preparatory meetings shaped how I shared the project in class in the initial sessions on campus, in which I explained the background to the project, the history of the Civic, and the commitment we made in our working together through the class, its varied motivations and responsibilities. Ideally, this first layer made explicit the ethos, ethics, and sensibility of the partnership and the project.

In this particular year, we headed straight off on the second Wednesday of the semester to meet the Valhalla Park partners and to get a sense of the neighborhood. In this ambitious start, I aimed for the class to engage the "field" from the get-go and to rework immediately all the messy notions that operated in classroom discussions. I was confident in the process of orienting students in this first session. It was multilayered, but comfortable; this approach usually produced good results. In retrospect, at the end of a hard day the very assumption that the orientation would flow just as planned should have been my warning sign.

We maneuvered onto the bus, which was, of course, a little bit late and not quite where I expected it to be. These banalities, though, were normal, manageable; metaphorically I tucked them under my arm. On the bus, I counted heads. I did not yet know the class. Retrospectively, it struck me this approach was a little loose and risky. As a rule of thumb, it would be best to know who needed to be back on the bus when we were done at the end of the day.

We reached Valhalla Park. The group of students was large, filling the meeting room off the library to overflowing. Students put out a circle of chairs for our introductions, but there were so many of us in the room we ended up crammed into a hodgepodge of messy rows. We introduced ourselves. Gerty introduced the Civic partners and told her story, her personal struggles with eviction. The students were captivated for the most part, though clearly eager

Teaching and Learning   **87**

as well to get going, to have a look "out there." We headed out of the library off into the neighborhood as a group, a large one, trawling like a school of fish up Angela Street.

As we made our way around the neighborhood, one young woman irritatingly, constantly took photographs with a conspicuous big, long-lens camera. As she clicked photos and spoke loudly, my attention fully focused on her when I heard her exclaim, "A gangster!" I grimaced internally and wondered what my partners were feeling. In the meanwhile, setting up her "perfect photo," she had wandered off, separated from the group. Keeping an eye on her too, Gerty moved to round her up, a stray and obnoxious sheep, hovering, not knowing her well enough to tell her to stop. Gerty confessed later that she was worried that the student and her very big, loud, and visible camera would go amiss, the camera swiped, the student mugged.

Who was this student, I wondered? How did she miss our introduction, which emphasized respect and a consciousness that we were in this place on the backs of the legitimacy of our partners, who were respectable residents and activists in this neighborhood? She clearly had not attended the introductory class, where we explicitly discussed a protocol for taking photographs only with permission and after interviews.

"What is your name?" I asked. She replied that she was another student's friend, and "a student in the Photo School," an institution down the road from the University of Cape Town. She continued, "I am doing a photo-essay on this class and project, and I will be here every week." My hackles rose. I asked her to stop taking photographs, to join the group. She continued, a little more subtly. Unable to resolve this issue on the neighborhood street, we hashed it out instead on the bus back to campus, and, at the end of the afternoon, in my office, where I explained why she could not participate in the class and the project in this manner.

This orientation session was a moment where we were "at" the zoo, snapping photos of "exhibits" we had come to visit; and, at the same time, we were the zoo ourselves, a spectacle, unruly, herded, both entertained and despised, raised eyebrows following short skirts and the noisy entourage we collectively made. I was not the only one disturbed by these commotions. Other students were uncomfortable, stressed. Gerty was worried. Aunty Fadielah, a devout Muslim, was quietly unimpressed. The list went on. There was no time to engage everybody.

We headed back to campus, literally hot, utterly discomforted. Sweating and stressed I rushed to pick my daughter up from day care, following a tightly timed plan from Valhalla Park to campus on the bus, then back down the high-

88   Chapter Five

way to get her. I bumped fortuitously into two other parents, both friends and colleagues, also researchers. Clearly flustered and out of sorts, I blurted out the way the afternoon had unraveled and my worries about the tensions we had left behind us, for our partners, in this spectacle of a start to the project.

Could I, they suggested calmly, focus on it all tomorrow? Discuss it with Gerty and in class, where I might turn the crassness of the afternoon, its spectacle and blatant othering, into a decent discussion? A discussion built not on concepts as abstractions, but on the irritation and frustration of this first, supposedly orienting, session?

That night I spoke with Gerty, who responded, "Soph, you know, they think we are making money off this project if your students come clicking photos like that. People asked me on the road later: So, what are you charging for those photos, Aunty Gerty? Where are you selling them?" She explained how problematic it made her and our Civic partners look and added, "We have to do a lot of work to make this safe for you and your students. I really was worried that the girl was going to be robbed. Everybody will think that students carry that sort of equipment." In class the next day, I opened the discussion directly. Many students felt irate too; we were a spectacle.

This moment and its debate were productive. To address this crisis, six students—committed, concerned, and energetic—volunteered to meet me at lunchtime to figure out a more explicit protocol for our work with the Civic. From this conversation and process, we wrote out guidelines for how we could proceed as a group in the partnership work. This document outlined the principles for how we should be in Valhalla Park, to reflect our role as "collaborators," built on the ethos that underpinned our project. The guidelines aimed to help students reflect on their roles, directed them to pay attention to our partners, the partnership and its configuration, the neighborhood context, what they might put at risk. They addressed questions of legitimacies, our ethos, and the practicalities of how we worked in this partnership and in our projects. They aimed to help us engage with respect.

Engage with Respect, a Guide

The partnership, this course, and its curriculum required and demanded energy, commitment, enthusiasm, and total respect. Were we all on board? This was a question I asked literally, as we headed to and from Valhalla Park, and metaphorically, every session and week, as the projects unfolded. We had a project schedule, logistics, built into the regularity of the timing of our class sessions. Every Wednesday we arrived at 2 p.m., or thereabouts, and left

FIGURE 11. Interviewing together

promptly before 5 p.m. The schedule was banal but essential. It set the rhythm of the class and of our work in Valhalla Park.

But at this initial stage, our preparation in the classroom was about respect, not logistics or method. I ran this discussion carefully, offering it as an invitation to students, as an opening frame, the ethos of our partnership. I drew annually on our "guidelines for the research," the product of confusion and discomforts of that earlier problematic orientation session, where much went amiss.

The guide was a contract of sorts between the university and the neighborhood for this work. It articulated on paper the respect that had to be at the heart of the collaboration. We operated in this context as "learners," as novices in so many senses, new to the neighborhood and its residents, its languages, expressions, its varied norms and protocols. In most instances we were new to the process of doing research, learning to see and hear with newly developed eyes and ears. This initial discussion helped students consider themselves as a presence in our partners' terrain, that which we engaged when we left the classroom and moved out into the city. Our partners were responsible for us, for our safety, in very real and concrete ways. They also had their own roles, multiple ones. The guidelines offered a way to pay attention to and respect the daily hard work that our partners maintained to uphold their own legitimacy—as residents, as leaders of the Civic and community, as mothers and daughters, as neighbors—the identities and relationships that anchored them in this place. The guidelines shaped the ways in which we engaged and worked in and through the partnership.

The guide explicitly offered and challenged students to unpack the assumptions that we invariably carried with us, directing us to turn "whatever was obvious into a question." Nothing could be obvious; if it was obvious, it meant we had too easily slipped into "us" and "them," stuck in our own subtle and

Chapter Five

FIGURE 12. Guidelines for Research

Guidelines for Research

Drawn up by: Oliver Manjengwa, Nyasha Chamburuka, Alison Swartz, Lauren Renard, Teboho Mojaki, Nic Rosslee, and Sophie Oldfield (2007)

ENTRY POINTS INTO RESEARCH:

- What's obvious, turn into a question to unpack and explore it more.
- Whatever gets articulated as an "us" or "them," is not precise enough for our research needs; it needs to be further explored in relation to who is talking (their name, age, place, relationships etc.), and understood in terms of the negotiation/nature of the conversation taking place (for instance, as a formal interview, a casual conversation, something between . . .).
- In doing the research, we are aiming to understand somebody else's perspective: what they are saying and why, in its integrity and wholeness.
- In doing the research, we think carefully about where we are individually coming from and how it shapes what we see and don't see, and how we interpret these observations and experiences.
- We are learning how to do research—this is 80% of what the project is about.

**ENGAGEMENT WITH CIVIC ACTIVISTS AND VALHALLA PARK/
7DE LAAN/8STE LAAN "COMMUNITY" ON PROJECT:**

- Respect for them and their huge base of knowledge and experience of Valhalla Park and housing issues is the basis of the project and our working with them.
- We appreciate and work carefully through their relationships with residents as they are the basis on which we can do the research and they need to maintain these relationships in the future when our project is finished.
- Sensitivity is really important—in greeting, asking questions, conducting interviews, in being "in the field"—in Valhalla Park, in people's homes, in the street etc.

NEGOTIATING TAKING PHOTOGRAPHS:

- Ask permission.
- Explain what the photograph is for (your essay, the booklet, your own interest . . .).
- Ask person/people you're taking the photo of, if they would like particular things/people in the picture.
- Bring them a copy of the picture (one or two or more that they would enjoy themselves) when we come back to share the research.

---

sometimes crass assumptions. In practical and straightforward ways, it gave us the tools that we used, shaping how we navigated listening, asking questions, looking, and engaging. It shaped our politics and our epistemology by recognizing our partners' knowledge and situatedness, the solidarities in which we worked through and with them.

As researchers and as partners in this process, our challenge was to suspect and interrogate categories, to be precise, to document and reflect on context, conversation, the nuances that helped us make meaning in this process, that helped us carefully "inhabit" and sometimes rework—we hoped—"the differences between us." In this approach, we aimed to locate our ideas and reflections, to ground our geographies. This process helped us analyze urban concepts and the neighborhood, problematizing notions of informality, the legacies of segregation, the agency of ordinary residents, foundational building blocks that shaped our analysis and understanding of the city.

The guide was also practical and political. It specified a conversation for photographs, for explaining their purpose and use, a protocol for permission and bringing copies back to residents interviewed and engaged in the neighborhood. We brought this guide with us as we entered Valhalla Park to begin our orientation. From 2007 on, it shaped the discussion and focus of our initial classes. Every student received a copy, which we periodically returned to, discussed, and developed further as the course unfolded across the semester.

Questioning What We Know

On foot, with our partners, we immersed ourselves in the research process. In teams of two or three, we began the interviewing process, building from the survey and a simple set of questions. Teams started by interviewing their neighborhood partner, piloting the questions, a way to get to know each other. A month of interviewing formed the core of each project. This process guided research teams through the fieldwork, from observing to surveying and experimenting with interviewing. Initially basing the interviews on surveys, we then developed in-depth questions to build semistructured interviews. Week after week, with the content of the interview and conversation, we worked on campus and in the neighborhood sessions to deepen the research focus. In this way the questions became more precise; they shifted. We ended the interviewing with life histories, framed around the precise research questions each group developed. These methods were increasingly qualitative, a process through which to practice and develop skills and confidence in talking to people. The tasks were designed as a process in which we took context seriously, listened carefully, observed, discussed, questioned, and reflected.

An iterative back and forth was critical to our learning and engaging, to our ethos and approach to this work. Students reflected on it in written weekly journals, which I and an assistant postgraduate student read and commented on carefully. This back and forth was a means to engage individually, to check and support student reflection and learning. Each week my postgraduate research assistant met with our partners before the students and I arrived. This meeting was a chance to touch base, to discuss how the research was going, to share and discuss what the plan was for the day, to share the weekly guide for

Chapter Five

the research session. Our partners had a chance to raise questions and engage issues they were facing in their groups and in teaching students. In turn, my assistant shared the student discussions and reflections, sharing what might need navigating, which student in each team might need some support. These sessions also worked to reorient the research and teaching process in relation to what was going on in the neighborhood, the shifting and sometimes unpredictable dynamics that shaped the research work.

Every year and every class, there were a mix of reactions that reflected the class diversity, its focus, rhythms, its occasional eruptions. Students varied widely in their point of entry into this neighborhood context. Reading the weekly journals, written following the sessions in the neighborhood, was an insight into who was in class, who I was teaching. It was a means to engage individually, with care and rigor, across the research in the weeks that followed. Experiential entry points shaped positions from which students started the project work, which they could reflect on in the initial journal writing. These were steeped in the race, class, and gendered identities we all brought with us into the project, in the experiences that shaped us and our understanding, in the inequalities in which we all lived.

A second-year student from the greater Johannesburg area, for instance, wrote appreciatively, "My first impression of Valhalla Park was in the form of a happy realization, which was that this settlement looked very much like home, where I grew up. The township I grew up in is named Katlehong . . . It is very similar . . . I even found similarity in some undesirable aspects such as the repugnant smells that grace townships across South Africa." Not everyone exhibited this familiarity and confidence, the right this student felt to even name a smell as "repugnant." Another young man, for example, emphasized discomfort as he reflected on his feelings of nostalgia. "We had a group of kids following us happily and chatting to some of my other classmates, some of those kids were even getting handshakes from 'white' students. My heart was saddened by this . . . I felt what it was like to be the 'other' well-off person. I mean as a child I also remember running after nice cars and 'white' people or township visitors with my friends wishing we could be those people." Embodying the "university" in this project and space, he saw and felt himself critically and uncomfortably through their and his own eyes and experiences. He explained poignantly, "To be that person I must say is not as nice as I thought when I was still young."

In contrast, another student, who grew up in a township in Gauteng, stressed dismissively, "What's the issue here? I have seen worse. I have heard worse. I have lived in worse. In fact, on arriving . . . I was wondering exactly where the much talked about problems were." Another young man, who grew

Teaching and Learning **93**

up in a formerly segregated coloured area in Cape Town, not so far from the neighborhood, was not impressed at all. After hours, he worked as a summons server for the City of Cape Town Traffic Department, charged to deliver traffic fines and warrants of arrest in areas surrounding Valhalla Park. He made explicit in his journal that "I spend most of my working day in the company of a police officer who is stationed" in an area immediately adjacent. For this student, our research felt deeply problematic because the police officers he worked with classified Valhalla Park as a "high risk area," an area in which they would not deliver summons because the risks of retaliation on summons servers were too high. He commented in his journal, "We are generally hated by the public in these areas." After the first research session, he checked on his system at work on summonses in the area, noting that there were "147 unpaid speeding fines (ranging from R150 to R1,400) for 42 vehicles, 38 pending court summonses, and 22 arrest warrants for 9 people." In short, he did not welcome his position as a partner in this project.

For other students—in many cases, white—this first session produced disorientation of various sorts. In every class the partnership provided a first in-person engagement in a township for some students, which they had more often viewed at speed, through the window of a car flying down the N2 highway. For many it was a "shocker." One young man, a classically well-mannered product of the southern suburbs of Cape Town, wrote, "My first impression is a strange mixture of interest, inquisition, and absolute alienation. It was like entering into a new world, one that had existed on my doorstep . . . yet I had never encountered it's [sic] like before. I was part of the same city organism as they were but felt as different from them as if I were a visitor from the moon." For some, this dissonance proved discomforting and difficult. Another very privileged student stressed honestly, "Truthfully, I do not enjoy doing fieldwork, I don't enjoy the objectification of people. Here I am an educated, white male doing research in a predominantly black low-income neighborhood, and the fact that we have to leave all our valuables back at UCT makes me wonder why we didn't choose a different research project." He reconsidered this position at the end of this journal, openly admitting that "I guess the reason I feel uneasy is also because at the end of the session I come home and it's another world where I feel safe and comfortable."

In contrast, another young man relished this first immersion. He had been desperate to escape, in his words, "the dreary, tiresome and often mundane process of regurgitating theory," his summing up of his university education to date. Instead, he wrote, "at last, a chance to move away from the monotony of UCT's lecture venue, and towards the unpredictability of the physical

Chapter Five

And Valhalla Park was most definitely the real world—a true product of South Africa's sordid past, a working model, an actual community constrained by the many forces at play responsible for the forms of poverty characterized by the evening news." At the end of the semester, he explained, "Our experience in Valhalla Park not only opened my rich white boy eyes to the world my parents had so vehemently and effectively kept me from seeing, but also allowed me to experience something no other UCT course offered—a chance to actually apply what I had spent so much time learning." A hope that he then tailored and pared down: "Sure, we were all beginners, but regardless of that fact, we were all given the chance to meet people whom otherwise we would never have even contemplated."

Our partners were crucial in the student engaging and learning. They helped students navigate questions and reshape their thinking from these initial starting points. Working in the areas of the neighborhood designated by Gerty and myself, research teams conducted interviews with residents and households, a process that brought challenges and opportunities. Here, the partner introduced the team to the person being interviewed, shared the purpose of the research. They helped students figure out how to negotiate interviews, introducing them to protocols for asking to speak to residents and how to enter and be in homes. It was in the interviewing that each team engaged with the lived experience of families in the neighborhood, experiences that are always personal and consequential. Mrs. Jooste, for instance, one student reflected in her journal, "was a mother of six children who were grown up with families of their own. She had allowed two of her children to erect in her backyard for their families, but the other four were still without In journals, as well as transcripts, students processed interviews, conversations, interviews, and research experiences in the days research sessions. Another student wrote with concern and admiration Mrs. Abrahams was retrenched at age 40 and lived off less than" another reflected on "a family supporting seven children off income." Journals and transcripts were a productive space to up, to reflect on the deeply personal nature of what the research made visible: neighborhood struggles that city statistics and effaced. Experiences were embedded in context, specific, many productive ways, they were truths, experiences means to engage the lived realities of its inequalities. ing process that students also worked through and research, and its politics. Most residents, for instance, outsiders, suspicious. A student commented on this reality as "slightly skeptic [*sic*] looks [that] often turned into smiles, a small twinkle in the eyes, a willingness to cooperate" when the interview was facilitated by Eric, his Valhalla Park partner. Nonneighborhood people in the area were few and far between, often city officials of some sort. A young woman student reflected on her community partner Aziza and her consistent "reassurance to households that we were in no way associated with the city council." She reflected that "people seemed threatened by us, they thought perhaps that their homes were somehow going to be jeopardized." Negotiating this turf was not easy. Another student wrote in his journal, "A young man of about 25 years old (relatively well dressed and wearing jewelry) crossed the road in front of us. As he passed by, Mina, our research partner, asked him for a cigarette. He shook his head and turned to face us, and then he just stood and stared. His manner was very masculine, territorial, intimidating, as if to say: 'what are you doing here?' That—at least—was my interpretation." Mina, he added, moved him and his group along quickly.

Through immersing ourselves on foot, in person, in conversation, in teams, we questioned what we knew. We tracked our thinking in the research rhythms and conversations of each group, with our partners, with residents, in journals. We learned to read the landscape and to listen carefully. We paid attention to the layers, the multiple registers, to the knowledge each offered and situated. The learning was rich and complex, observed in everyday interactions and in interviews. It was tacit and felt, in the reluctance and joy, in those moments of hesitation, of openness and hospitality. It was negotiated in the varied ways people sometimes welcomed research teams into homes and at other times left them outside on the other side of the gate. These crucial nuances were bound up in the partnership. We learned, we read, and we shared. Our partners situated this work in their knowledge, immediate and proximate, lived, immersed in this context, in this home, on that street, in that history and its politics. We engaged slowly, incrementally, carefully. This pedagogy helped us question what we knew, guiding an iterative process, a way to learn and interrogate patterns and meanings of inequality and difference, a way to interrogate our project and ourselves.

In Homes, Not Shacks! Interrogating Readings

In a clear voice, indignant, the student's words rang out across Studio 5, our large classroom space in my department. "These are bungalows; homes, not shacks! This fact, the literature just failed to get." She was frustrated with the literature, its partial slicing of the everyday complex realities we were explor-

Chapter Five

ing. Her analysis traced the dignity in the backyard family's story, the reality of the family's housing. Her team described the home, its layers and materials. They portrayed this housing option as a dignified urban form—a bungalow, a reading that ran counter to backyard "shacks" as "just put together," remnants discarded by others, invisible, a purely makeshift urbanism. The student was positive, hopeful: a city can be so much more than its parts, the structural analysis of inequality and exclusion. Ordinary people can create change. Our partners create change every day.

My approach in this pedagogy intentionally intertwined the practice of research with reading and writing to find a point of critique carefully, that incision point in which to build theory. Consistent writing made this incremental pedagogy possible.

To deepen our engagement methodologically and analytically, we drew, for instance, on literature in which researchers shared field stories and experimentation with methods and their limits. In the interviewing weeks, I twinned together these articles with a prompt to explore the research experience and process that week. Students had a choice of prompts, questions that challenged them to link their own field experience together with the journal article, to think beyond what they saw, to conceptualize and to work between their experience and the literature and its notions of method and rigor. This interweaving of reflection on our practice through engaging with literature deepened our analysis. It thickened our reflection and helped us fine-tune methods.

The course culminated in students drawing on the weekly journal, field notes, and interview transcripts to build an empirical story for the final academic paper. This layering of different forms of writing aimed to help students build rich and precise findings. Students drew from the interviews and life histories they completed in their research teams, from observations and conversations, which proved a thick and critical foundation for writing papers and building arguments, a key goal in our curriculum.

In this research process, many students changed how they thought. They embraced the partnership's relational mode, its epistemic challenges, its substantive richness and realities. A young woman student, for instance, recalled in her final journal,

> Today I spoke to one of the cleaners in the Microbiology building at UCT. I simply asked if she was not home for the holiday, and she ended up telling me her life story. I feel that sitting and talking to the residents of Valhalla Park has changed me a little bit. I feel I've learnt to listen attentively, and I have also

Teaching and Learning **97**

seen the very basic need of people to feel that they are being listened to. I do not know if I would have spoken to the kind lady today if I had not done this research in VP. I think that the course has affected me more deeply than many others I have done in the past . . . I would not call myself ignorant, but it can be very easy to sometimes turn a blind eye to other people's suffering, and to think that the whole world lives in the way in which you do.

In and through this partnership, likewise, I changed the ways I taught.

A Toolbox for Writing

The pedagogy provided a space for students to experiment with writing, building techniques for developing thick description from field notes and interview transcriptions to narratives and papers. In writing journals and field notes, students navigated the research process, stepping back, reflecting, then returning to ask more questions to engage further. Moving back and forth between the university and neighborhood built a form of accounting and sharing. The pedagogy of writing, its layers and iterative quality, fostered the layers of analysis, their substantiation. The depth of analytical questions extended, situated in an urban studies literature and rooted in conversation, in the substance of what people shared, in the research process in the neighborhood.

Back on campus, I worked with students, helping them find ways to thread the richness of interviews and field notes with their own experiential reflections and participant observation. What could be made of this mix? What were the varied ways these elements could be linked together? We worked together to develop the empirical threads in the research to weave it into a "story," something that had a beginning, middle, and end. In helping students experiment with writing thickly, I encouraged them to share the complex struggles that shaped the neighborhood and city. I loved this part of the process. It took a certain type of energy and rigor. It took confidence and trust in the process, as well as good advice and teaching on my part. This form of writing enabled students to move between the stories and transcripts from neighborhood interviews, debates in the city, and the literature we drew on in class. This movement built a rigorous relational approach to theorizing.

I encouraged students to locate and enrich what they found expressed in the literature, in urban geography and theory, in ways that opened rather than reduced everyday lives. This was the heart of the challenge I posed: to build urban geography in and from the ethos of our project, to do justice to Valhalla Park, to the rich stories that were told to us, to the people who told them. I

FIGURE 13. Reams of journals and reports

APPENDIX

Analysing how households in
Valhalla Park sustain a living

Mea
Rifqa

EGS 315 S
URBAN GEOGRAPHY
Group Project

Valhalla Park:
Negotiating Backyard Living

PREPARED BY:
Is'haaq Akoon
Tzvetomira Kirova
Katherine Mann

29 October 2004

worked closely one-on-one on the outlines of these empirical stories. Dani's eyes sparked, for instance; a small smile crept across her face. She got it! She could see, feel even, the story she wanted to tell, the argument that was emerging as we discussed her fieldwork, as we brainstormed on paper. She had a plan, a vision for the paper. I loved this creative moment, which helped students nurture the story from their interviews and find their own voice through it. Rich and rigorous qualitative analysis built on this thick description. It brought the reader into the context carefully. It described the place, the people, the feel; in other words, yes, that living room context was critical, the old man's passion in telling you this; certainly, the shopkeeper's weariness in explaining how this all worked, absolutely; and, yes, the passion and the place of your partner and your research team in this was key, too. In this process, theory could be fashioned, engendered, and crafted in particulars.

Students worked hard to produce research for their coursework papers and for the multiple publications that were part of every project. Close engagements with me in this process helped develop these materials, the findings and argument, in thick writing, in relation to the relevant literatures that we engaged. In the home-based businesses and the informal economy project, for instance, did you draw on the messy realities of family business strategies to complicate Christian Rogerson's sharp distinction between survivalist and growth-orientated enterprises (1996)? Could you build your analysis on the stories of women-headed households and their struggles to put food on the table? Are you mirroring Deborah James's (2012) notions of popular rather than informal economies, which seemed so evident in the family and street economies in the project? What was missing in this scholarly argument and in that paper? Threads of family history pride, and disappointment, that emerged from interviews on neighborhood businesses? Yes, there is the weft of your argument. With these prompts, students began to imagine placing themselves in conversation with literature and theory, making a shift from relying on others' arguments to carefully nurturing, then asserting and building their own work in a research conversation in and beyond Valhalla Park.

Writing worked in small, incremental steps, in the layers of research notes and journals, through the poster, a draft outline, then a paper. A hundred words here, a hundred there; shifts in thinking accumulated and mattered. Analysis built from the complexity of the stories of renters, landlords, and shack dwellers, and of our partners and their families. Our writing was situated, reflective of the research practice, located in yards, on stoeps, by the front gate, behind burglar bars, on the street corner, in land occupations, on the field, in the neighborhood's self-built homes.

We built a writing practice in which the research process, its method, conversation, dialogue, and our learning was visible. It was substantively rich and specific, infused in complex realities, in homes, learned through people's lives, their predilections. It was deeply theoretical. The partnership held us steady to question what we knew in grounded and relational ways. It generated a body of theory that was a part of our process, its complicities and productive discomforts, its power struggles. This epistemology stretched across the partnership. It shifted the ways in which we wrote, how we accounted for and developed the work, what we knew and how, with whom we knew and why.

Critique Leavened with Love

Studio 5 was full of people, the partners and their kids, mothers, friends, students; the buzz filled the hallways, tumbling over into the atrium below. It was a different sound than normal, not the rhythm of a lecture, a single voice, semidroning above the ambient noise, not the normal banter of a laboratory or practical session. Instead, there was a jumble, a cacophony of tones and accents, a few babies crying, some little children running at full speed in the hallway outside the classroom upstairs. A full Jammie Shuttle bus, rather than a *kombi* taxi, had gone to Valhalla Park to pick up our partners. They had asked earlier if others could come too. And I had said yes, of course, our only limitation is the forty-four seats in the bus. All forty-four seats, and then some, were full.

Our partners had come for the end-of-semester presentations of my students' final papers for the course, built on our research project. This paper was the capstone piece of writing in the curriculum. The students presented it as a final part of their mark for the course. Each year our partners joined us on campus for this session, in which we assessed the students' work together. This was an important moment for the students, for the course, and for me, a key marker of what we had learned and produced. It was a moment to return the layers of weekly hospitality and assessment that encased our project and our work together all semester long.

The campus visit was a big occasion. The first year in which we held this session, our partners dressed up, arriving awash in shades of pink, pastel, subtle, soft, and vibrant. Hair beautiful, turned out. We took a picture to mark the occasion. I felt embarrassed in the normal jeans and top I had pulled on unthinkingly that morning. For Valhalla Park participants, coming to campus meant many things: it was an event, a moment to celebrate our hard work in the partnership, a moment in which we recognized their experience and knowledge. For many, it was a rare chance to come to a privileged part of

Chapter Five

our city as invited, recognized, and celebrated guests. In an interview halfway through our decade of work together, Rosemary, for instance, reflected on her shock that first year coming to campus. When I asked why she was shocked, she explained, "Look, I brace myself when I leave Valhalla Park to go into that part of the city. I brace myself to be humiliated, and to be put in my place. I was shocked because I came to campus and I was welcomed, I was loved, I was treated like a VIP by my students."

All went well until a dreadful presentation, in which racist stereotypes ran amuck, were said out loud under the guise of the research project. I squirmed in my seat. Right behind me there were four Valhalla Park women. I thought I felt their eyes burning into me, particularly as this young man explained "that women in Valhalla Park have children just to access the child grant." Oh yes, of course, the then-R280 a month would set you up to live like a queen, I thought. My stomach churned—what rubbish was he talking? I could not even dream of passing him. The basics of the project were missing, the process of looking, listening, substantiating.

I tried to think on my feet. I invited Faranaaz, his research partner, to comment. Would she like to say anything first, I asked? She jumped up: yes, enthusiastically she turned to the student and to the class. "Yes, that is my story, that is the truth," she exclaimed.

This was a moment of truth for me, too. A moment of dissonance and truth, a clash of the supposed rigors of methodology and "university" ways of knowing, with her truth, her hard-won, hard-fought, everyday experience, her way of making sense of herself and the world. I recast in my head my questions and comments for him. I also adjusted my mark. It was still a terrible presentation and a shoddy piece of work, but if it was her truth and story, he certainly could not fail.

My discomfort in this instance reflected a knee-jerk, standard academic critique: avoid essentialism at all costs; avoid naturalizing what are socially constructed notions and categories. In this instance, it was not useful. It was not enough. Much more productive was an overarching principle in our pedagogy to "take another look," the principle that underpinned our aspiration to unpick and unpack our assumption making, our theory building. On reflection, this suggestion might have been excellent for the student as well as for Faranaaz, a suggestion for all the sides of our partnership in this moment.

Our projects were full of this back-and-forth dissonance. Instances like this one were peppered across the student presentations. They often arose for our partners, when students critically engaged with issues of gangs, gangsters, and violence in Valhalla Park. They arose for me in assessment.

In the spirit of partnership, I invited my partners to assess the students' final presentations. I aimed to place my partners' assessments parallel to my marking and commenting, my filling in of a presentation-grading template. I constructed a partner evaluation form, a very open-ended document with a short, and I hoped welcoming, note at the top of the page: "Please give your students comments on the presentations and your work together. Thank you!!!" The remainder of the page was left blank, with most of the page an open space for our partners to write their comments on the presentations. I included a small space at the bottom of the page for a mark, but the form didn't emphasize this element. It was clearly optional.

I expected comments on the presentation, on the content, I expected, I now realize, some mirror of my own set of comments, my own criteria for a "good presentation," for evidence of "good" research and hard thought and work. Not all the partners chose to write. But those who did complete assessments rarely mentioned the analysis, the substance and stories that had been told in the narration of our research.

The forms, instead, overflowed with assessment of the experience of working with each student, of being with them in the partners' homes and neighborhood. The writers commented almost without fail on the student, on her personality, his generosity, her care and attention to them and to others in the neighborhood. In short, love jumped off these pages. Care sprang with it, heart, and emotions. A sharing on paper of what it meant to work together on the project, a mix of admiration and respect.

To say I was struck by the contrast between what I had expected and what emerged is an understatement. I smiled and had to take my own advice: *Take another look.* I had assumed a very particular notion and target of critique and assessment focused on the presentation, its structure, substance, argument, and the way it was conveyed and organized. These were the criteria that shaped my and the university's notion of academic rigor. In contrast, our partners had celebrated on these forms the students' way of being with them in their homes and neighborhood, the embodiment and relationality of the research practice in this course. The partner assessments highlighted so clearly what the course had demanded and what underlay each presentation, what it might or might not have offered in its analysis. Comments did not draw on a language of deficit, of what the student had not done, or what she could have done better. Instead, the assessments were love letters that expressed so much of the partnership and its process.

I copied each assessment together with my own presentation comments and handed them back to the students. The assessment by the research part-

FIGURE 14. Contrasting assessments

Please give your students comment on their presentations and your work together.

Thank you!!

The presentations was for me a pleasure to be at.
I did hear a lot of different stories that have
happend with people in my community. It was
a big shock to hear the terrible things that they
had to face in there past.

It was a great experience to work with
Ann Sophe & Will. I had the chance to work with
some of the other students and I am very happy
to got the opportunity becouse I did learn
a lot during the reseorch. This was my first rescorch
and it was spetacular, and I hope it will not
be my lost. This few weeks was excited to work
with the students, It made me feel different than
be for I begon to reseorch. I wont to help
in my community in the future. I have seen a
different side of the people in Valhalla Park that
I lived with all these years.

From

JAMIELA

EGS 3015S Project Marks
(essay 50%, appendix 15%, presentation 10%, individual journals 10%, group and self assessment 15%)

Vimbai & Megan — In your final mark, I will take into consideration that Eva dropped out ½ way through the project.

Group Name: Vimbai & Megan

(69) Report: Overall — you've woven together an interesting argument which you've contextualised quite well see following pts/ paragraphs.
- in general, any literature references need careful contextualisation — eg. Govern — if this really relevant here?
- the utility theory needed much more substantiation —
Is it appropriate for a qualitative research analysis —
be careful in "mixing" — to matching method & theory!
- make sure you, in future, reference interviews
- graphs derived from your data work well (eg. Figure 2)
- Figure 3&4 - additional contextual detail about inside & outside of home would make this graphs more useful.

Appendix: (Note length max - 10 pages!)

(64) You've included lots of material — but it's inconsistently presented and organized. The bullet point interview descriptions are not v. useful. A bit more effort in this section would have been good — see instructions — inclusion of map, overall table of data etc...too.

(70) Presentation: You both presented well with a coherent structure & plan. You drew together a wide range of materials in the presentation & you tried to engage with some wider theoretical issues. I liked the way you contested the assumptions about overcrowding by showing benefits it can derive & also the social relationships that are central.

With the quantitative materials, you need to be careful about the qualitative — challenge central to the research process — make sure you have a match between method & research questions before, & substantiation. *Megan — your references didn't show much at all — thought you actually presented quite well.*

NB: be careful about generalisations re: "classed areas" & people & their ways in your narratives.

Thanks for your enthusiastic and committed engagement to the project!

I'll email your individual marks once all your self & group assessments have been submitted

[signature]

106   Chapter Five

ner and mine jarred and, at the same time, worked together. They could not be reconciled. They did not need to be wiped away. Instead, our differences opened reflection. Our process offered a space to acknowledge and inhabit difference, to work to let these types of dissonance be productive, a space or moment through which we might rework our approach and understanding.

It was precisely this mix of critique and love, academic rigor and relational embodiment, that made this partnership a compass for radical pedagogy. It made visible what was gained and what was lost in a particular mode of critique, and what might be reworked and rebuilt in combining critique and care.

In this pedagogy, students found a voice, an energy, a purpose, a way of writing; some found a passion for the field, for urban studies methods, for its questions, for city politics. The approach was methodological and epistemological. It was also ethical and political, shaping the ethos of the practice and purpose of our research. It was—for some—also ontological, a way of being in the university and academy, in the city.

Coda—Teaching

Teaching repositioned expertise. It was our compass, essential for producing and nurturing the next generation of urbanists. I was one of many teachers. Partners were teachers. Students, partners, and I were learners. We shifted university practice. We worked in the disjuncture of our city, its segregation. What was normally the purview of the professor, of the university—the curriculum, its standards, and norms—was stretched and challenged. This approach to teaching brought together and juxtaposed the rigor of social science and the rigor of neighborhood partner expertise, debate, and logic. Teaching and learning thickened in these relationships, in their rigor, care, and epistemology.

In this way, the project of urban studies was intimate and concrete, embodied and located, conceptually complex, rooted in the trajectories of our city and its inequities. In this back and forth, we addressed the disjuncture of our field, its bifurcation in literatures and policies, in a dislocated view on the city. This passion and logic were the modus operandi that shaped the teaching through the partnership. At stake was the nature of our urban studies syllabi and curricula, pedagogical practices refigured in partnership.

Out of this new pedagogy, we built an archive of research, a wide array of publications for varied publics across the city.

CHAPTER **6**

Research

A Web of Writing Practices and Publics

Writing Practices

The partnership produced a wide variety of genres of research publication, from research posters, popular books, maps, and a *Yellow Pages* directory to academic articles, student theses, and this book. Publications were diverse in form, in their origins, purpose, and intent, produced in a range of writing practices and for diversely situated publics, in the plural.

Through publications, we navigated writing. Unlike the research interviewing process, anchored squarely in the neighborhood, writing was driven more by the university, through student writing on the research, our class and its pedagogies, and my own scholarly work. We experimented with writing and genre. We inserted it in varied ways into our research process. Research posters, for instance, became part of our process, to share analysis early on, to build a conversation and to account for our interviewing work. They were a genre that we brought back to the neighborhood, to our partners and those individuals and families interviewed. They aimed to invite conversation and debate immediately after the research was finished. We developed popular books to share narratives, to highlight lived experiences we researched. Sometimes an issue required and demanded harder data, such as maps, which more systemically visualized an issue or debate like housing access or settlement building.

Some genres recurred annually, such as posters and maps; these lay at the heart of the partnership and its process of working and researching. Some publications were distinct, unique, spinning out in specific ways, singular and narrow. Some stretched, in their inclusiveness. They wove thickly and dura-

Chapter Six

bly, existing beyond a particular project. Others were soft and fragile, threaded more thinly.

In this chapter, I share the stories of producing these genres of publication, their logics, and possibilities. The politics and trajectories of publications, for instance, were sometimes contradictory. A map, seemingly neutral and objective, could quickly be drawn on to tell a story and to make an argument. A journal article, seemingly a scholarly privilege, could ground and land elsewhere. Our *Yellow Pages*, intended for local use, could easily be caught up in wider city politics.

In examining these genres and their interplay, I reflect on the web of publics, politics, and the slices of "city" varied publications brought to view. I explore when, why, and for whom genres of publication proved critical. I consider the ways in which different types of publications enabled ways to account, engage, and review our research work. Here my intent is to juxtapose, not reify, the differences between forms of writing, grounded in their particular geographies and in the partnership and its collaborative method and approach. In working in and between conventional categorizations of popular and political or scholarly and theoretical forms, the partnership publications stretched notions of research writing and expanded forms of urban theorizing.

Research Posters Taped to Walls

The research posters were displayed, big and bulky, visual and vivid. They were taped to the walls of the neighborhood crèche in which we held the research party for the project on home-based businesses and making ends meet, scheduled three-quarters of the way through the semester. The posters aimed to invite conversation and debate, to share early findings with our partners and those individuals and families interviewed.

We started the proceedings, the semiformal part of the event: a moment to thank our partners and the families and individuals interviewed. Our neighborhood partners and students offered their thoughts and words. For instance, Lefien reflected on the research presented on the posters on the walls around us. Her words in Afrikaans held the room. She spoke powerfully about the posters, their reflection on the neighborhood economy, about everybody's efforts to stay afloat, about unemployment and its ills. Her analysis reflected our project and her knowledge accrued in years of trade union activism, her present work as a piece-rate seamstress, working evenings in a garage in Athlone, about five kilometers away. After Lefien, students shared their thoughts and experiences. Residents acknowledged and teased their favorites, calling on them to speak.

The public nature of the posters was important and provocative. They were by design engageable, readable, intended to be recognizable. The stories they included were direct from interviews, with photos of homes and families, of varied neighborhood contexts. Families could read their own stories, and their neighbors', see themselves and others. Unlike less accessible journal articles and books, the posters attracted all sorts of readers and an immediate set of comments and discussion. They also showed the making of the research, its method, and location in particular interviews and homes: on Polar Street, as opposed to Joanna Street, at Aunty Eleanor's home as opposed to Fatima's house.

We introduced research posters to share the research more effectively. The sharing was key. Each story had to stand up to scrutiny in the neighborhood. The poster presentation held the research teams, the students and our partners, to account because those we interviewed came to read our work and to participate in the discussion and celebration. The research poster needed to portray the reality it engaged, what it meant to be living in Sewende Laan, the struggles of home-based businesses, or the passions of Klopse, for example. The public conversations generated through the posters shaped the careful ways in which we translated and brought individual interviews and life histories into analysis. It grounded the ways in which we mapped and shared photographs. They were an accessible and open way to develop ideas, to test and share emerging arguments. They produced a rigor, one that was immersed in the neighborhood as well as in the university.

Conventionally a conference device for showcasing mainstream academic papers, the poster form was transformed in this context. Its purpose was refigured to share our process and to test the ways in which this work resonated with those with whom it was produced. Posters challenged students because they required visual conceptualization, as well as narrative and analytical strategies. The success or failure of our posters built on the ways in which the research was recognizable, resonant, the ways it did or did not speak to truths, everyday practices, hardships in some cases, celebrations in others. The posters and their emergent stories aimed to reflect the nuance and contingency of the ways in which we had listened, asked, and engaged. They challenged students to account for the knowledge, to situate it in context, to locate it. The poster, as a genre, and the research party, as an event at which we shared it, extended the rigor of our work. Its public engagement put center stage the pol-

FIGURE 15. Research posters

112 Chapter Six

itics of how and for whom we produced this knowledge, to whom we had to account.

"There We Are on the Map"

They were looking hard, examining, tracing their fingers down Long Street. This family was gathered around the map of Agste Laan, the newest and most controversial informal settlement or land occupation in the neighborhood. All women, two older, three younger, a baby in arms—I moved a little closer to join them. "Look, there we are. That's our place, our door. See us, I am in the picture." For this family, finding oneself on the map was a moment of profound recognition, a small but concrete form of acknowledgement. This moment was precious for our project too, a glimpse into what it might mean for a "shack dweller" in a relatively new and insecure settlement to find her house on the map. It was these layers of identification, of territory, of home building in the face of so many odds that the map shared and legitimated. It was its readability, its legibility that marked its difference and helped families find themselves, literally see themselves on this map, in the settlement.

We were in the big Nooitgedacht Community Hall, posters and maps tacked to the dark turquoise walls. The Agste Laan map hung in the center of the longer wall, huge and a little unconventional. Each research team had contributed a representation of their research area. These sections of the map were pieced together, a multiscaled jigsaw, a complex task. The map marked and celebrated Agste Laan's first birthday, its survival despite city policy, the odds stacked against it and the families who had relocated here. The map recorded and reflected the extraordinary work of the settlement's making, the piece-by-piece construction of homes, the everyday work of building and settling. Its form disrupted convention, piecing together photographs, hand drawings, textures of homes, and streets, which created a thick, layered representation of the settlement. In this collage form the map recorded families finding a space in the city, inserting themselves into the neighborhood, being present, building homes and security incrementally and precariously, here and now.

The map was also the culmination of the research project, an aggregation of the work of all the research teams across Agste Laan. Research teams were asked to record plots, according to their sizes, and homes, their shapes and building forms, the conversations they had with home builders and with settlement families. Research teams had opted to complete this research work in various ways. Many had the home and plot marked, a picture of the home layered on top. In some cases, the photos recorded, and in some cases, celebrated, the home itself. Some homes called out for this attention, for instance, the beautiful pink house on Bubblegum Street, the home with the immaculate brickwork going into the off-street parking and *stoep* on Long Street, and the white and black polka dot fencing marking the garden of a house built closer to Modderdam Road. Other research teams chose to include a photograph of the family in the house, the residents themselves. These photos spoke for themselves, often carefully composed, families standing together, some with smiles, proud, others more ambiguous.

The map was also much more than these individual pieces and layers. It blew open a narrow notion of research and data as an expert-driven process. It disrupted a Cartesian, god's-eye view. It was a map, at scale; it was also intimate, a pieced-together product of our research, a portal into our process, a rich source of knowledge. It brought into its topography photographs of families, standing outside homes, working on roofs, building walls, making this settlement a place to call home. It gave shape and texture to the work of home building, tracked in the photographs, in the brickwork, in the spacing and organizing of this informal settlement. It demonstrated the messy texturing and building of the community, homes constructed on top of an old netball court, the reconstituting of municipal fencing around a wetland as the settlement encroached and swallowed it. It hinted at the networks and relationships connecting settlement families. It was relational in a secondary sense as well, a product of our partnership and this research. We were in and on it, as were our partners too, those who lived in the settlement, as residents and as researchers. They were a critical part of this map and knowledge making.

Maps were central to most of our research, a genre we could share, which our partners could use in their organizing and mobilizing. Our maps were also part of our method, a way to track our interviewing and systematize our research, and a form of analysis, a record of neighborhood spatial patterns and dynamics. They were a particularly important publication genre in our projects on housing insecurity, in the Agste Laan and Sewende Laan research, and in the backyard projects. In these contexts, the maps marked as present and significant families that were technically, officially, and politically landless and illegal. The map visually shared this critical work to self-construct homes, to find and develop shelter, to not only claim land but to make a place.

I drew on slivers of the stories of families living in this informal settlement in several scholarly articles. What happened to the settlement's form and content in this shift in genre?

FIGURE 16. The Agste Laan map on the wall

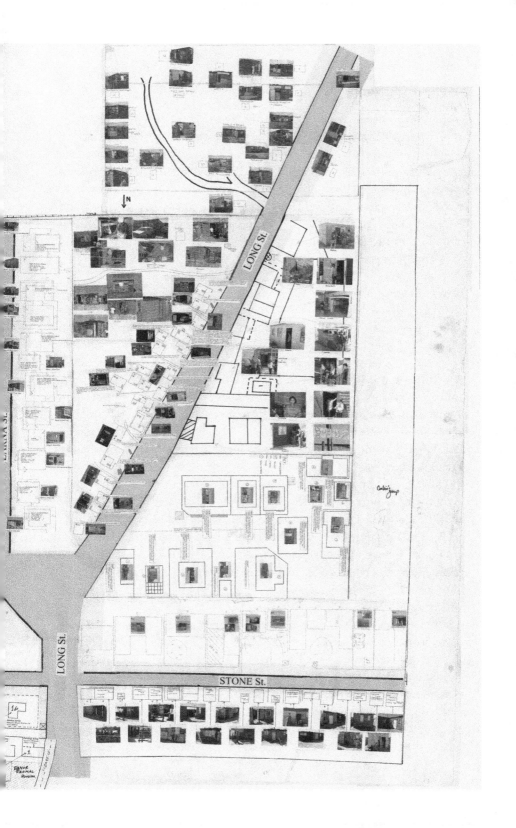

116   Chapter Six

Located in Journal Articles

In some ways journal articles sat far from the worlds of everyday life in this neighborhood. They were built in sets of engagements with colleagues, assembled in varied university-linked discussions. Each published piece was a slice of the partnership's work, separated out and distilled in the frame and focus of the journal in which it was published, in the preciseness of its scholarly conversation, dimensions, and written conventions. In the logics of writing and publishing journal articles, the partnership and the everyday neighborhood world of its practice were rendered visible and made digestible in partial and always-specific ways.

Different strands of work shaped my writing for journals from the partnership. They evolved in particular ways in published finished form. In a jointly authored piece, for example, my colleague Charlotte Lemanski engaged with elite gated communities, the focus of her research at the time, and I reflected on city responses to land occupation from the partnership work (Lemanski and Oldfield 2009). The comparative kernel for the article emerged out of a meeting in Paris on "Territorialization in Cities of the South," a conversation between French, South African, and Indian researchers. In contrast, in a book chapter written with my colleague Kristian Stokke focused on practices and politics of neoliberalism, I drew on fine-grained narratives of the anti-eviction and anti–cost recovery strategies of the Civic, inspired by Gerty's and George's leadership. Although collaboration with Kristian originated in a workshop on the politics of democracy and decentralization, this chapter was published as part of a collection, *Contesting Neoliberalism: Urban Frontiers* (2007), edited by Helga Leitner, Jamie Peck, and Eric Sheppard, geographers intent on building a global comparative conversation to challenge narrow notions of neoliberalism and its city politics and governance effects.

The trajectories of the papers were, of course, specific. They were shaped by the research materials, but the argument of each paper evolved in the specific scholarly conversation, the genealogies of past published work we drew on, and its dimensions and vocabularies. Charlotte and I, for instance, suggested the need for more careful interpretation of the contextual and relational logics of local processes such as gating and invading, as well as a more precise assessment of the state's engagement with these urban development processes in increasingly southern cities like Cape Town. The article's published form in *Environment and Planning A* led to a particular register of theory, rooted in each case but framed in a language and literature of state-society encounters in a body of work on southern cities. In the chapter with Kristian, we moved between fine-grained narratives of activism and social theory that challenged monotone accounts of neoliberal city politics. These dual threads connected the chapter across the collected volume *Contesting Neoliberalism*, which tracked powerful acts of city making between and across the global south and global north, making visible the tensions and practices that shape neoliberalism in practice and in the plural. My parts of the article and book chapter were products of the partnership work, but little of it was evident in the final form of either piece. The visible slivers of the partnership were determined by the arguments and the conventions of the journal and the edited book in which we published the work.

By their nature, accredited articles and scholarly book chapters are a particular kind of urban genre, a specific form of theory, bound up in a scholarly review process. Each paper was submitted and underwent a rigorous review process, an anonymous set of assessments. In the book chapter, driven by the edited design, our piece was interrogated on its own terms and placed in conversation precisely with the project of the volume. In the article on land occupations, reviewers challenged us to situate more accurately the comparison at a conceptual scale. The reviewers were not overly concerned or engaged with our rendering of the gating and occupation practices but asked us to be much clearer in the analytical work embedded in our comparison. The most critical reviewer could not stomach our comparing such opposites: land invasions with gated communities, although it was the essence and point of our argument and paper. We edited, revised, and polished, defending our selective engagement to the editors in writing, drawing on the language of the journal and its forms of review and critique.

This protocol for publishing was a politics of academic submission, a form of accounting and vetting steeped in a particular type of hierarchically organized scholarly public. It was steered by editors, their networks of and negotiations with reviewers, as well as their assessment of our work, their imagination of the publication audience and conversation. In each case, the peer-review process pushed us to edit and refine, to carefully layer, polish, and make each argument precise. Eventually, each piece was finalized, officially vetted, proofed, and copyedited. Each entered the virtual world of online journal downloading and hard copy and e-form, in select libraries, shaped by academic library subscriptions and the geographies of purchasing possibilities and limits.

This academic review process and its politics were bound up, on the one hand, in broader processes of professional development, career building, research funding, and the like. On the other, they were anchored in some form

118 Chapter Six

FIGURE 17. Deciphering the practice

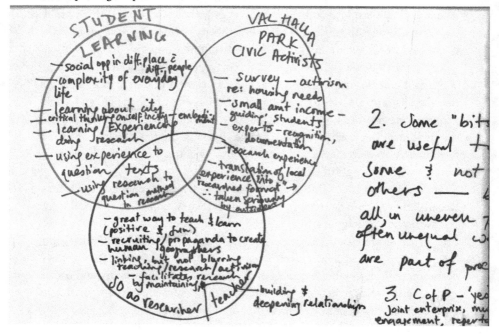

of networked anonymity, explicit in its "double blindness" in which authors and reviewers remained anonymous. Conversation and accounting were channeled through the editors and their discretion to accept our responses. This review process was so different from those that we accounted to in the neighborhood in research parties and through popular books. These contrasting accounts were public, in person, intimate, a sharing of neighborhood experiences and stories. They shared and assessed our capacities to put experiences into words and arguments, to assess what was at stake, to make it visual and visceral in photos and on maps on the walls of the neighborhood hall and its crèche.

In select ways, the worlds of the neighborhood and the partnership were made visible in my scholarly publishing, part of a register and record, part of not only our project archive but a broader urban debate, woven together. While presented at best as exemplary and bracketed, in partial form, the textured everyday realities of the neighborhood nonetheless entered the world of scholarly urban debate. In other ways, each published article and book chapter sat far from the worlds and lives of families in the neighborhood and in its settlement, the contexts that inspired my own thinking and contributions to

Research **119**

my published work. In stark contrast, partnership publications were designed precisely for the neighborhood.

Yellow Pages in Every Household

The *Yellow Pages*, a neighborhood business directory, for instance, was completed for the project on making ends meet. Yet, the public in this publication was not straightforward or simple, either. Despite our intent to publish something useful for the neighborhood directly, the *Yellow Pages* did not reduce neatly to this narrow notion.

We printed fifteen hundred copies of the neighborhood business directory, which our partners dropped off door-by-door across Valhalla Park. The *Yellow Pages* was a simple but effective record. It documented all the businesses that we mapped and interviewed in 2012. It was the result of rigorous and informed research, a systematic assessment of the neighborhood, a process in which we had combed each street, interviewed those businesses marked through signage, and found those less visible, small enterprises in homes, in the neighborhood's nooks and crannies. Families often knew about the places on their block, across the street, the larger establishments: the fish shop, the bakery, and a convenience store, all formal establishments on Angela Road, the main thoroughfare. But others were invisible to those other than neighbors, to those not in the know. The photographers and videographers, the spice shops, the informal fish suppliers, the bakers and sewing outfits, the small ways that families experimented to make ends meet, to earn a little, and in some instances, to keep long-standing family traditions alive in the form of these businesses. It was these small initiatives, some signposted, others not at all, that the research documented and the *Yellow Pages* shared. Only by immersing ourselves in the neighborhood could we become "in the know," our collective research objective.

The research project was completed amid neighborhood and citywide xenophobic violence (discussed in chapter 4 of this book). Could we still publish the *Yellow Pages* in this polarized and violent context? Worriedly, I asked Gerty this question prior to its printing. The research had not focused explicitly on xenophobia, but this emotive and violent politics shaped any publishing of research on home-based or small businesses in this moment. This was an important question, reflective of the complexities of producing alternative research outputs. Gerty responded to my question clearly and directly. We must publish the *Yellow Pages*. It was even more important to do so in this context. Publishing it was part of our accountability as a partnership, a collective

FIGURE 18. Our *Yellow Pages*

BARBERS

Rayyaan Marcus's Barber Shop
Rayyaan Marcus
58 Matteus Street
073 802 2326
Downtown hair cutting services

Zahir's Barber Shop
Zahir
103 Angela Street
073 0677 504
8am-5pm

FRUIT, VEG & MEAT

Boy's Fruit & Veg
Victor
95 Matteus Street
021 934 0228
6am-7pm

Green Shop
Ibrahim Laurence
32 Paulus Street
083 716 7917
8am-7pm

Jerome's Fruit & Veg
Fruit, veg & wood Jerome
Rosenberg
27 Joan Street
083 5899 233
8am-5pm

Mario's Fruit & Veg
Mario
174 Angela Street
076 119 0391
9am-7pm

Madi's Fruit & Veg
Fruit, veg, spices, stationery
Donavan Madumbo
195 Angela Street
9am-5pm

Veronica's Fruit & Veg
Veronica
102 Eleanor Street
073 537 5394
7am-10pm

Merle's Meat
Meat
Charles
62 Johannes Street
072 436 1542
Weekends

Patrick's Fishing
Fresh fish
Patrick Jonas
10 Edmund Street
072 093 6378
Flexible

Gammie's Fish
Fish, fruit & veg
Azizi
47 Rosalind Street
021 934 4925
10am-6.50pm

Poor Man's Friend
Fruit & veg
Shamiela & Toys
4 Edward Street
021 9048 873
8am-late

Poor Man's Friend Stall
Fresh fruit & veg
Hilton Louis
Cnr Angela Strt & Charl's Lane
073 290 5632
8.30am-7pm

Fieka's Tuckshop
Meat & veg
Fieka
V25 7th Laan
071 2922 344
8 till late

122 Chapter Six

decision, configured so differently from a conventional framing of this question in an unpartnered, academic context.

I moved forward with the printing. An agent at Top Copy printers helped. He gave us a good price for our fifteen hundred copies when he learned of our intent. He figured out a pocket-size version, small yet locatable. He brought the *Yellow Pages* alive by redesigning our cover, bringing the man known as the Pastor, a local leader and businessman, to the forefront, a visual invitation to open the directory. It was in this very politicized context, following the forced removal of Somali and most other "foreign" traders, that this simple document made its way across the area. A presence, not a solution, it marked the end of our project, the purpose of our interviews, our going door-to-door.

The *Yellow Pages* did not directly address xenophobia and the attendant politics and violence. Instead, it situated it, bringing to the fore layers of complex issues in which xenophobia operated and erupted. These included, for instance, the histories of neighborhood businesses, their successes and longevity, as well as failures; the specific patterns and practices of gangsterism that shaped neighborhood streets and access to business spaces; and the reality and hardships that sustained and provided a logic to the networks that interlinked the neighborhood and broader city informal economy. The *Yellow Pages* was distributed to all the households in the neighborhood. It prompted us to rework our thinking. On the one hand, it showed a neighborhood hard at work, eking out a living. On the other, its shadows revealed xenophobic tensions, stories of competition, and the feelings that drove that xenophobia. Through its recording of the businesses burned out, run out of the neighborhood, their presence remained, if in that moment only on paper.

As an alternative genre of publication organized explicitly for the neighborhood, the *Yellow Pages* sat in this visceral, violent politics, its contestation of the right to work, to be in this neighborhood and others like it across the city and nation. Immersed in questions of identity and violence, the *Yellow Pages* made visible a layer of the meta-politics of the city and the research itself, its purpose, its intent, its publics and politics.

"That's My Book!"

Gerty laughed as she recalled the telephone call from a friend from Belhar, a neighborhood five or so kilometers from Valhalla Park. "Did you see that book about you, about Valhalla Park?" the friend asked. Gerty retorted, "I told her: That is *my* book, I wrote that book!" *The Valhalla Park Community Entertainers: Klopse Building a Better Neighbourhood* was in demand. The two hundred

Research   **123**

copies printed flew out of Gerty's front door, into the hands of residents, the troupe, the Klopse Board, and the councilor.

For both the projects on Klopse and Sewende Laan, we developed the stories and experiences for publication in short-book form, a genre to share the research, to highlight lived experiences and struggles, to ensure these struggles and hard everyday realities reached the public domain. Both books documented the Civic's work and mobilization, sharing the intimate practices that made Klopse happen and that sustained families in the Sewende Laan informal settlement. These short books worked as a mode of writing and critique that could contribute to the Civic's work, to its activism, to forms of neighborhood building. Popular, political, and advocatory, they offered alternative stories and accounts from our research that rubbed up against and challenged a discourse of deficit and lack—of informality and marginality—that so often characterized the neighborhood and its place in debates in the literature and the city.

The Klopse book, for instance, was beautiful, warm, alive, sharing page after page, story after story, and photographs of proud residents, the performers, singers, costume makers, the cooks behind the scenes, the leaders engineering the logistics and coordinating the buses and uniforms, the directors and the coaches leading biweekly practices throughout much of the year. The text highlighted the commitment and passion of Gerty and other Klopse leaders, many of whom were our research partners too. The book shared the hard "engineering" and organizing work that underlay Klopse and the neighborhood troupe's participation year after year. It made clear its effects, the ways it built community in the neighborhood, in the yearlong neighborhood fundraising and coordinating, and training of the minstrels, young and old, and in its pinnacle, the performance in the city center and neighborhood at the New Year, and in the annual summer competition season.

The book substantiated Klopse as the glue that sustained and legitimated the Civic, as an organization, and its leaders, as representative of this neighborhood, year after year. Klopse brought a wide array of residents into the minstrel troupe, and through it, into the Civic. Young mothers met older activists, not just as neighbors but also as leaders of the troupe. They enrolled their young children as *trompoppies* (drum majorettes), marching decked out in Klopse gear, a key part of every troupe. Children participated in the youth choir, attending weekend practices for months on end. Mothers came along, listening at first, joining later, as they were increasingly woven into the fabric of the Civic and its work. Young men spoke of their pride in the Klopse band, in their roles as trombonist and percussionist, in the music, in being part of

FIGURE 19. The Klopse Book

THE VALHALLA COMMUNITY ENTERTAINERS:
KLOPSE BUILDING A BETTER NEIGHBOURHOOD

"The community is so much stronger now than it was before. The area just feels friendlier, more like a real neighbourhood, instead of just a bunch of people who happen to live on the same streets."

FATIMA MCKAY
FRIENDLY COMPETITION

At 46 Polar Street is Fatima who as soon as she learns about the nature of our visit, lights up and springs into action. She pulls drums and tambourines from on top of a closet, modeling them and demonstrating how they are used, explaining who in the troupe uses them, where they are positioned in the march.

Fatima runs into the other room and calls over her shoulder, telling us to begin asking our questions while she unearths at least six-years' worth of Coon costumes from a plastic bag. They trail behind her as she re-emerges into the living room and she answers our queries through a sea of glitter and sequins and shiny fabric.

Despite her long-standing membership in Elsie's River, she still feels a strong connection to the Valhalla

Chapter Six

this family and community tradition. Some participants pictured in the book explained how through Klopse they had sidestepped the temptations of alcohol, even in some cases the agonies of *tik* addiction and gangsterism, struggles they saw friends face. This type of integrating and connecting wove through the Klopse book as a publication.

For both the Civic and the troupe, the book made clear to the world "who we are and what we are: proud and disciplined." As noted in chapter 3, Gerty explained this crisply in the book's introduction:

> Although we were more than successful in stopping the problem of fighting between gangs and drug trafficking in our area [and we fought for and won our housing project], we still faced a stigma. When people hear of Valhalla Park, they associated our area with the 28 Gang and drugs. We needed to do something to show the world out there and the people out there—standing in Cape Town, across the Cape Flats, in the surrounding areas—to show them here we are. We come from Valhalla Park. This is what we are doing in Valhalla Park. And this is what we can do. (2010)

It was this motivation that had led to the establishment of a minstrel troupe in 2005.

For both the troupe and the Civic, as well as for the Klopse Board in which they competed, the book worked to counter a broader politics in the city, in which Klopse was contested, understood as disorganized, chaotic, as a working-class, "coloured" practice and performance, problematically racialized, representative of a racist history. As Mrs. Kamalie explained in the book, "It's my history, my mother's, my grandparents'—it's part of me, part of the Cape, part of slavery" (2010). Klopse Boards argued that Klopse and its city-wide celebration was developmental, that it worked to build community. This argument was discounted, even ridiculed by the city and by politicians, especially in budget processes for funding the Klopse New Year celebrations and in the bureaucratic processes to get permission to compete in stadia around the Cape Flats. It was this politics with which the book resonated, sharing Klopse as activism, a practice that forged crucial bonds that held this neighborhood and this city together.

"The Story of Sewende Laan Is like a Book"

Similarly, *My 7de Laan* shared a David-and-Goliath story, the story of a group of families living in an informal settlement and the Civic, an organization with no resources that contested the city's intent to destroy the settlement. This

Research   **127**

short book tracked the struggle for security that marked Sewende Laan residents' accounts of their lives, the meaningfulness of building and fighting for the right to remain in this settlement. This was an alternate story of city building that documented the hard-fought ways in which the nearly one hundred families involved defended themselves against the city's attempts to demolish the settlement. The book included each family's story and photographs of them and their homes.

Scattered across its pages were small-scale plans of the settlement's self-built homes, the piece-by-piece evidence of families building Sewende Laan. The book mapped the settlement's struggles and achievements, the stories of its streets and families, which were not recorded either officially in the city's maps, or physically with literal street signs. George Rosenberg Avenue and Gertrude Square Street, for instance, paid homage to Gerty and to George, now sadly deceased, as Civic leaders and as key leaders in the Sewende Laan struggle.

In the introduction to the book, Gerty framed the land occupation and its defense as part of the Civic's work, its commitment to supporting families struggling with shelter. Several partners, for instance, lived in Sewende Laan: Dan and Lefien, while Mina migrated in that period between her shack in Sewende Laan and her mother's house, her kids living with their grandmother in the winter, and then, over time, all year long.

We distributed the book to the residents of Sewende Laan, to other neighbors, to the Legal Resources Centre (LRC), the NGO that defended the settlement in court. They requested twenty copies of the book to distribute to funders and to others in similar struggles elsewhere in the country. The book stood in part as testimony to the meaningfulness of this NGO's work. Beyond the precise legal judgment and the work in court, the stories in the booklet spoke to what the NGO's defense meant in everyday terms to this settlement, to the Civic, and to Valhalla Park. It shared the relief and the joy families felt in their hard-won security, being legally allowed to persist and build their lives in the settlement. It gave intimate, personal meaning to the broad-brush constitutional rights that the NGO defended and for which it and the Civic fought. It documented a combination of sacrifice and hard work, of great success and its significance. The book recorded as well the paradox that despite the court ruling that formal housing, services, and legal title must be granted by the city, Sewende Laan families persisted, waiting still for the court judgment to be made material in the form of formal housing.

I reflected on both books, on what was political and advocatory in them, and on their popularity, their meaningfulness. They resonated with and shook

FIGURE 20. *My Sewende Laan*

130 Chapter Six

up my own writing on urban politics, challenging my narrower notions of what constituted scholarly narration and argument. In the resonance of the Klopse stories, for instance, it struck me how singular, blinkered even, my own thinking on the Civic and its activism had been, focused, like the literature I am in conversation with, on material struggles, on services, on a narrow set of issues driven by a particular notion of radical politics. While spectacular, often hugely successful, even epic, these threads of activism traced in the literature, and in my own writing, were partial. They were limited in both a language of development, in a practice of policy that was city driven, and in narrow scholarly notions of progressive activism and its politics. At the same time, the Klopse story disrupted the phenomenon's narrow conceptualization, reworking common sense notions of its legacy, its politics, its place in the city today. The stories in the book challenged its reduction to a racial stereotype, a tourist spectacle, a middle-class embarrassment as an apartheid hangover. As Gerty claimed and demonstrated, against common sense, an activist could be a minstrel and a minstrel an activist.

As a genre, these short books traveled in and across the neighborhood, the city, and beyond. They had ripple effects, offering resonant stories of activism for land and housing, for identity in Klopse, for a place in city politics. This mode of writing and critique—in partnership—met celebration and performance, a positive neighborhood building. In their meeting, in their rubbing up against each other, we could rethink developmental categories. We could find ways to challenge and rework notions of dysfunction that stigmatized the neighborhood and our partners, categories that so easily laced urban debates in a language of exception, dysfunction, and dystopia. We unsettled and conceptually enriched notions of "politics" and its subjectivities and city practices, the political terrain in which the partnership operated.

Spinning Off, Student Research

Alex, my student, and Suki, our partner, teamed up to explore the ways in which local informal businesses link with citywide and, in some cases, transnational trade networks. Francis worked with Mina, Gerty's daughter, in Agste Laan, documenting young women's bodily insecurities in the settlement; Rifqah, with Uncle Dan, continued interviewing families about state grants; Evan worked with Lefien to explore neighborhood women's histories of work in and retrenchments from Cape Town's textile industries. In different periods, Simone joined up with Gerty to research water cutoffs, Inge to explore histories of eviction. Siân and Saskia developed their master's thesis research through the partnership, working with Gerty and me.

Well-thumbed housing waiting list letters shared by residents in the research project on overcrowded neighborhood rental housing inspired Saskia, for instance, to build on the partnership for her master's thesis. In interviews, residents shared these letters. They pulled them out of a box or a pile, or found them neatly ordered in a drawer or cupboard. As third-year students in the project on overcrowded public housing, she and her research partners had tracked and traced this evidence from household to household in the project on overcrowded rental housing. The research group started the process of documenting the ways young and grown-up "children" and older residents alike prioritized signing up on state waiting lists to access state-built housing, even though access in the short and long term was so unlikely. The idea incubated and a few years later, Saskia registered for a master's degree under my supervision to develop this research on waiting for housing further. In Siân's case, during her work as a postgraduate assistant for the partnership, she was struck by conversations with the few residents who had purchased homes in the neighborhood, through a government policy option to promote home ownership and sell off city rental stock, something rarely mentioned in our work on housing. I approached Gerty and our partners to include these thesis projects as extensions of our partnership.

There was something beautifully generative about these links between student thesis research and the partnership work, which proved a space to find and then imagine research, to cultivate an issue, to frame it in its dynamics, its importance, its situatedness in the neighborhood. Waiting for housing, for instance, was taken for granted, uncommented on. Over lifetimes, and across generations, public housing in the neighborhood had become almost uniformly overcrowded, housing grandparents, parents, grown children, and their children. Perhaps not surprisingly, the majority of those over eighteen were registered as waiting for homes from the city. But what did younger generations expect when they signed up for housing? Did they expect to receive a home? What did they do in the "meanwhile," while they waited for years, sometimes decades, to move into their own homes? In contrast, the choice to purchase a home in the neighborhood was quite rare. What motivated those families? What were their logics and investments, their engagement with the city, their expectations of the Civic? Gerty helped Saskia and Siân, respectively, develop these research questions.

The situating of the project in the partnership provided a place to arrive

132 Chapter Six

and depart, literally in Gerty's living room and kitchen. The legitimacy of this student research was encased in Gerty's presence, her status as a Civic leader, and our long-term commitment to our research partnership. It figuratively framed the research in a far longer conversation, one that exceeded the bounds of a thesis project. This thick set of relationships shaped how interviewees responded and engaged. It outlined the expectations for this research, linking it to the Civic's mobilization for homes and the organization's claims and encounters with the city. It situated the work in a rigorous set of expectations that the work would be relevant and rooted in the neighborhood.

Yet, a thesis is not conventionally a collaborative collective project. Its syntax and its form are organized individually, assessed by examiners; the protocols of research are individually driven, or at least asserted as such. A clear and well-developed thesis must be disciplined, rigorous, framed in and conceived through engagement with scholarly literatures, by a discipline and its knowledge and writing norms. The thesis work and writing were conceived and undertaken as part of an offshoot of the partnership, its long-term work, its forms of accounting. Students stressed how hard they worked because of our partners and because the research engaged concrete, taken-for-granted lived realities in the neighborhood.

An Archive across the City

I found the large and small maps, which had been displayed at meetings, bundled together in a tube in the corner of my office, reams of student papers on the bookshelf, and student theses stashed in a drawer. Documents, reports, papers, photographs, maps, stories, data, recordings, and transcripts of conversations, life and family histories were our partnership's archive, a vast and varied mix produced over the decade. Files and files of student journals chronicled the research process, its learning, field notes as companions. These varied documents sat side-by-side with folders of newspaper clippings, piles of articles, policy documents, and city data, with the web of publications we generated. Laced in memories, sights, sounds, the inspirations and joys of our collaboration, its energy and enthusiasm, the archive lived on in our pedagogy, the modes of our collaboration, our learning and confidence in our research capacities. This was our partnership archive.

In search of our early research findings, less cataloged and copied, I dropped by several partners' homes to ask if they still had copies of their group research reports. These visits were a chance to reminisce and visit, to catch up after a long while. I found Washiela at home. Following her parents' deaths,

FIGURE 21. Scrutinizing the posters

she had stopped participating in our projects, taken up with the care of her and her sister's children. Washiela pulled out her group's report. She emerged quickly with it, a reflection of her newly renovated and well-ordered home. Fatima found hers under the bed. In passing, Masnoena said she would have a good look. Zaaida found hers and Gerty's in a box, stashed in the lounge unit, at the bottom of a box of documents, a marker of the years that had passed.

As I gathered and pieced together these varied materials, this archive, I was struck by the fact that—a decade later—the reports were intact, remembered, held onto. Shelved across the city, publications were dispersed in cupboards in Valhalla Park, in my office, elsewhere at UCT, in my home. The *Yellow Pages*

134 Chapter Six

we had produced were spread across neighborhood households. Klopse booklets were prized possessions, hung onto by troupe captains and directors; they had made their way into board members' pockets and cupboards elsewhere in Cape Town. The Klopse posters, laminated so they would survive, were slipped behind Gerty's wardrobe in her bedroom, pulled out at troupe meetings in the years following the research project. The Sewende Laan book sat in the Legal Resources Centre's offices in town, a few copies off elsewhere in the country. Made up of many pieces, the archive stretched in and between the neighborhood, the university, and beyond.

An Interwoven Web

We lost and we gained in these interwoven writing choices and strategies, in a book and its investment in celebrating a right to occupy land; in an article's commitment to a right to the city, an elite and popular politics; in our *Yellow Pages* and its politics of building the neighborhood economy. We spun in and beyond the neighborhood, spinning a web of publications and publics. Some parts of this web were networked, interlaced together. Other parts were entangled, knotted tightly in place. Publications wove together, plural, layered, and multidimensional. They gave the web shape and scope. Some were durable and long lasting, others faded with time, ephemeral, short lived. Some were intimate, proximate, others distant; some fragile, others bold. Some were distinct, unique, interconnecting momentarily, spinning out in specific ways, singular and narrow. Some stretched and included, were receptive and adaptive. They wove together thickly, existing beyond a particular project. In this web, we built an embodied and living archive across the decade, constituted in the partnership and its varied publications and publics.

Publications had varied origins, purposes, and intents. Some genres, such as posters and maps, recurred annually; they sat at the heart of the partnership and its process of working and researching. They sustained our partnership, beyond projects and semester-long courses. They showed the limits of our work, as well. Some forms of research and writing became provocations to act; others were put aside. Some found shape in a neighborhood and city discourse, and some spun through the geopolitics of urban theory, its rooting and reshaping in often-southern city experiences. In each a politics was at play, sometimes fraught, at other times positive; sometimes powerful, at other times latent and ambivalent.

We experimented with varied strategies to improve and deepen the outputs of the research, ways to share project outcomes and effects. These ideas reflected careful planning, the inspiration of each project and its emergent arguments, as well as the intuitiveness and contingencies of the partnership work. We experimented to push further, to follow our hunches, to pursue ideas, to push boundaries. And sometimes we searched in the dark. We navigated the complexities of everyday life, of city making, its strategy and dissonance, the hurt, material deprivation, and violence, the generosity and care. The partnership located this work, in literal, epistemological, social, and emotional terms. It was epistemological, collective, accountable, and embodied, forged in the relationships through which we built and sustained the partnership.

Rooted in the partnership, its processes and projects, the partnership archive was varied. It challenged narrow definitions of knowledge making. Publications did not materialize in a single type or form, in a bounded or narrow notion of "theory." In extending and connecting genres and publics, each publication was an intervention that was concrete and specific, sometimes powerful in its simplicity. Each publication, its stories, form, and questions helped rejig our imagination of how we might theorize. Across this web, what worked and for whom, when, and where exceeded any simplistic notion of what might be useful or popular, academic or scholarly.

Coda—Publications

The partnership's archive unsettles and disrupts a narrow notion of what counts as research in academia. In its geographies, the archive linked us, working across the topography of publics to whom we accounted, with whom we engaged, through whom we hoped and worked to change our city. The partnership produced a wider, thicker, rigorous mix of publication—from mapping and directories to pamphlets and books, to process documents and teaching materials, to journal articles and this monograph. In its varieties of forms and uses, it offers a wider, rooted, and located body of work, a flourishing of form and genre, of intent and use, of publics and audiences, of impacts and registers. In the archive's varied forms and uses, publications produced in partnership extend and deepen narrow conventional notions of academic research. In this frame of work, theorizing in partnership challenges conventional university notions of what counts as research. In doing urban studies differently, the partnership makes problematic the dominance of a narrow form of scholarly publication. In collaborative work, what is at stake is the edifice of academia, its capacity to respond, to engage, to account, to take root, and to find relevance across the city.

CHAPTER 7

Theorizing the City Otherwise

In Stories of Collaboration

I shared the partnership practice in stories of collaboration, shaped by each project, its tempo, cadence, its feel, and vocabulary. In backyarding we found our feet. Ambitiously systematic, we pushed ourselves to figure out a way to complete the project, to survey all the backyards in the neighborhood.

Sewende Laan felt special, a chance to document an almost impossible victory. Agste Laan brought new territory, its layers complex, its pace contested, dissonant. In our celebration of Klopse, we found a tempo, improvised, inspired in the neighborhood's passion, pride, the hustle to make it happen.

Punctuated by challenges of making ends meet, the home-based business project immersed us in business closures and xenophobia.

The Civic project was diffuse, fractured, like the myriad issues that shape this neighborhood, the demands pulling it in a hundred different directions.

The narratives make the collaboration and its rhythms tactile, visceral, felt. The partnership's rhythm was syncopated: in the tempo, timing, and language of the community, its activism; in the tempo, timing, and language of the university, its curriculum. Side by side, syncopated rhythms shaped our teaching, the research outputs, our mode of working and being together. It drove the ways we improvised, worked with contradictions and conflict, and compromised.

In stories of collaboration, I wrote this rhythm, its unexpected beats, its multiple tempos and contingencies, its dissonances and harmonies. The narratives make the rhythm of the partnership and its participants visible: the partners, myself, my students, neighborhood residents. They show the ways in which, through the partnership over time, we built a way of working together, an embodied form of practice and knowledge.

Narratives invite readers into the partnership, into its thick context, its intricacies, its costs, its complicities, and its inspirations. Stories share the day-to-day work of teaching and researching together, the small, mundane, banal elements. Stories track teaching and assessment that moved in and between the university classroom and township neighborhood, in and out of ordinary people's households; through these movements, the stories track the changes in learning—of students, of research partners, of myself. Narratives share the contingencies of our partnership, our feelings and hopes, our forms of critique, the epistemologies and ethos that sustained our work.

Narratives lay bare the feel, the emotion, the visceral elements that shaped how we worked together. The writing is laced with affect, with the emotional labor through which we threaded the partnership together over the decade. Stories share a range of emotions, from joy and pride, to love, to occasional anger and fear, to inspiration, to the care and generosity that sustained the partnership. Across this terrain, narratives share the pleasure of the relevance of this form of partnership, its learning, and its dimensions.

The partnership grew as we grew with it, as others joined and left, as issues to work on shifted, as our politics and preoccupations moved in time. The stories are themselves political. I chose stories that show the work, the commitment and the perseverance, the patience as well as inspiration that led us to keep going year after year, for a decade. The stories describe the varied risks and stakes, and the ways we worked through them. They are epistemological, in their repositioning of the work of teaching and research, and of the community and university in it.

Stories bring into view our collaborative practice across its contingencies, in its fullness. They share the process, the organizing, the strategizing, the evaluating, the thinking on our feet, the long-term planning. I share the fun, the humor, the organized and the unexpected pleasures, what made us happy, what we loved, what inspired us to return to our work together, again and again.

The stories share the weight of our purpose. They describe the seriousness with which we took our research, our chance to document and engage. They unearth the everyday struggles, their layers and histories of injustice, the politics of the possible wrought through the Civic's commitment to activism, and the slog of everyday life. Stories share mistakes, compromises, complicities, as well as surprises, humor, love, and care. I recalled stories where scholarly notions of rigor, critique, and progress were turned upside down. I chose stories that shared articulations of justice and legitimacy, moments when the possibilities and the contradictions of our city were laid bare.

138   Chapter Seven

In these narratives I reveal moments when we shifted our assumptions, moments that challenged us, moments that brought us together in unexpected ways, the resonance and discomforts felt when we confronted critical questions. The narratives disrupt categorical notions of the neighborhood and university. They confirm ideas and reshape them, thickly textured in content and layers. They reveal the contingent contradictions that shaped our work together, the paradoxes that entangled us across the city. The stories reflect choices, thinking, and learning, rooted in the partnership work, rooted in the city.

In this interpretative form, stories make the world of the partnership visible. They juxtapose the difference it embodied. In narrative technique, stories thread competing ideas and contentions loosely together. In these juxtapositions, the narratives show ways we inhabited difference, and sometimes reworked it.

Stories offer an epistemic break with characteristic scholarly writing. They are a form of "truth-telling." They situate claims and partialities, rendering commitments readable, located. They show the ways in which the partnership made it possible to shift thinking and theorizing. They articulate conflicts and compromises. They are written to resound with the urgency of partners' truths, to juxtapose and position my truths. All of this *is* truth-telling: truth-telling in ordinary words.

In Ordinary Words

The partnership stories, its archive, are infused with ordinary words. They are a product of these contingencies of choices, of context, of thinking and doing. They are a product of each project, its practice. In the partnership's practices, over time, ordinary words became more than the sum of their parts. They became conceptual tools. They had a genealogy, transparent in our process, visible through the stories of our collaboration. My words were in English, a reflection of the hard limits of my Afrikaans, the circuits of my knowledge and (in)capacities. My partners had other words, vocabularies in *Kaaps*, a Cape Town creole, in Afrikaans, and in English. These ordinary words emerged in the partnership work, simple, and powerful.

- *Dignity* found in, and fought for, in homes, in organizing, in the Civic's work.
- The endless search for work, the struggle *to make ends meet, to put a pot on the table.*

Theorizing the City Otherwise   **139**

- *"Not once, not twice, but thrice,"* the imperative to stay vigilant, to defend and claim rights and resources, to struggle in and against the city.
- *Community work,* "only finished when there's someone else behind you *to carry on."*
- *Family* and its power to sustain and shape, to authorize in households, across generations.
- *Pride,* deep-seated in the neighborhood, in its struggles and its history.
- *Violence,* an everyday household pain, a jagged neighborhood edge, a structured city reality, an epistemic stigma.
- *Compromise,* in the contradictions and conflicts in which we collaborated and were complicit.
- *Community,* in everyday knocks on the door, in the contingencies of crises, in the work of building belonging and insisting on justice.
- A *partnership,* an invitation across the city, the legacies and structures, the languages that divided us.

The words, and the phrases in which they appeared, were pedagogical and conceptual. They reflected context, their location in the contradictions and struggles of each project, its focus, in precise interviews and intimate conversations. They reflected the expertise and experience of the partners, the debate so evident in projects and their conflicts. They were reviewed and engaged, tested, and discussed, assessed in ways that ensured they held traction. The partnership's ordinary words, and the stories through which they emerged, reveal their location, a genesis in the partnership's biography, the genealogy of our work together.

Ordinary words also articulate the partnership's mode, its practice, and methodology. They developed from the partnership's practices, its ethos, its epistemology. These vocabularies emerged in the substance and rhythm of the collaboration. These words were collective rather than singular, twinned together.

The partnership entwined us,

> back and forth,
> incremental and long term.

We learned and documented between neighborhood and university.

Words emerged in the writing, in the genres of publications, in the publics, in our archive; in substantiating the partnership. They were

Chapter Seven

> rooted and mobile,
> > durable and fragile.

They track narrative pathways, written in the stories, developed from the partnership's practices, its ethos, that which allowed us to build relationships and sustain them with

> rigor and respect,
> > trust and compromise.

In the expertise and experience of our partners, in struggles for justice, for recognition, we built a way of teaching and researching together, through

> struggle and care,
> > critique and love.

The partnership offers this vocabulary as methodology and epistemology. It was the literal and conceptual means by which we were capable of travel, across the city, across a decade. I brought these ordinary words back to the university, to the classroom, to my scholarship. They became part of the rhythm of my thinking and theorizing.

In the research party at the end of one project, Raksha held up a bright red page. On it in bold black marker she had written the words, "insightful, inspiring!" She explained, "Valhalla Park, you have inspired me." As she sat down, another student jumped up. He held up a blue page. It read "perseverance." He called out, "You have taught me how important it is not to give up." Other students joined in, a cascade of words: courage, honesty, hardship, joyful. Each word shone, simple, an affirmation of struggle, the hard work, the graft of our partners in the neighborhood day after day.

The letters that partners wrote to students were equally full and meaningful. Zaaida commented that her student partners were "humble"; she appreciated that. Aunty Meisie had, she wrote, taken her student partners "as my own." She remarked on their respect, which she would keep in her heart. Shireen appreciated "expressing her feelings" with her partners. Jamiela remarked that the work together in her team "made her feel different than before," aware of a different side of the people "I have lived with all these years," wanting "to help my community in future." These words proved a means to share and to account.

They mark the ways we inhabited the partnership, as

- residents, neighbors, students, learners, and researchers;
- activists, minstrels, fans, struggle plumbers, parents, and leaders;

Theorizing the City Otherwise **141**

- timekeepers, caterers, logisticians, guides, and assessors; and, as
- friends, a professor, researchers, experts, partners.

Thick in its practice, the partnership was intimate, concrete, a substantive way in which to work together. Ordinary words trace the pathways we found to work together, a product of the partnership's pedagogy, its process, its contradictions. Grounded and located, ordinary words are responsive to and reflective of the partnership and its life worlds.

In Verbs—In the "Doing Words" of Practice

Verbs formed a key set of these ordinary words.

- We *mixed and matched* questions and imperatives, partners and students.
- We *unraveled and fused* commitments and concepts, publications and research outputs.
- We *juxtaposed and entwined* the languages and commitments of the neighborhood and the classroom, community and university forms of expertise and accounting.

The verbs articulate the doing of the partnership practice. Ordinary verbs did not become tools overnight. They thickened over time in the partnership practice, in the trust that we built. In partnership, they became theoretical, condensed in their productive tensions, embodied as ways of working together and of theorizing.

IN TEACHING

Ordinary verbs emerged in teaching.

We mixed questions and imperatives for research. We matched that which we were invited to engage in the neighborhood. We fused together the work of research and teaching, the university classroom and the so-called field, the partners and students as research teams, processes of joint review and assessment. The teaching demanded precise unraveling, substantiation to delve deeper into *that* story, *that* argument. Our partners' questions reframed debates, showed other truths. Like students and me, partners paused to listen and hear, to engage, troubled and inspired themselves by stories, depths of hardships, strategies to overcome, the complexities and contradictions of everyday life. In this contrast, I saw and felt the rhythm of my thinking, my comments, my criteria for rigor, the particularity of my scholarly norms.

142 Chapter Seven

In teaching we unraveled our questions and presumptions. The logic of the partnership fused our positionalities as researchers and learners within it.

IN VARIED RESEARCH OUTCOMES

We entwined neighborhood norms and protocols with scholarly methods and research systematics, respecting these varied forms of authority and legitimacy that shaped the neighborhood and university contexts. In these juxtapositions, we lived with the dissonances, we examined them and inhabited them.

We shared our research work in the neighborhood and in presentation sessions in class on campus. It was an obligation located in the neighborhood, not a privilege of the university alone. In these moments of reviewing, we paused. We stopped to share and to make apparent the travel between interview and conversation, between a neighborhood struggle and the research. Our tempo shifted to check, to ask: Could the person interviewed recognize their story? Could they trace and track between the conversation in their home or front yard and what appeared in these posters, in publications? Were our partnership and our process visible in these written layers? In conversation with our partners, and with me, students reshaped questions, extended ideas, discussed critical tensions, and rewrote. We adjusted and reacted, changed plans. We found new rhythms and ways forward when conflict emerged, when our legitimacy was questioned, when the partnership needed extending, its form shifting.

In the layers of review, we built the publication process across the community and the university. The layers of assessment went beyond protocols of informed consent—consent and accounting built in practice, over a decade, and layered in our process. Consent and its accounting were substantive, checked at multiple points, on posters shared in research parties in the neighborhood, at presentation stage on campus, at the draft book stage when permissions to use stories and photos were garnered, in the returning of photos framed and narratives printed for individuals and families. This thick form of accounting entwined university and community protocols, the invitations, and the refusals. It juxtaposed rather than erased difference and disagreement.

In the creativity and push to publish genres of writing in a variety of registers for multiple publics, we built an archive, a body of work responsive to the publics to which the partnership accounted.

The archive became a living web in its research products: the posters, the books, the maps, the *Yellow Pages*, the papers, the student work, this book. These varied forms of writing were threaded and textured. They embodied struggle and care, critique and love.

Theorizing the City Otherwise   **143**

IN THE SLOW TIME OF A DECADE

The partnership work together was necessarily slow.

It built an ethic and ethos of care, in relationships embodied in the partnership, in growing trust, in our capacity and confidence, and our wish to work together. This trust grew through my relationship with Gerty and in her relationships with others in the Civic—neighbors and friends, her daughters, community workers, activists, stalwarts of this neighborhood. It built on the trust of my colleagues, those with whom I taught, who had faith in my running practical sessions, laboratories off campus, in this partnership. It built on my growing confidence to extend this work, to push it further, to experiment and commit to the partnership. It was shaped by the ways in which we worked together to teach students, to think again, to see in a different way. It was malleable, able to adapt when rhythms were interrupted, punctuated by the politics of the partnership, of the city that surrounded us.

IN THE RHYTHMS OF EMBODIED THEORIZING

Practices emerged in and from the rhythms of our relationships, from the rhythms of our work together, from the pulse and the tempo of projects, their accumulation and layering.

The theorizing built in the rhythms of activism, its ups and downs, its stops and starts, in the Civic's vigilance, its readiness to engage the city, to defend the neighborhood, to claim rights; in the imperative to get on with life in the meanwhile. It grew in the rhythms of university teaching, scheduled, regular, predictable, a weekly expectation, a curated syllabus, layered, a crescendo at the end of the semester. The theorizing evolved through the rhythm and structure of the partnership, its routine, its chronologies of events, conversations, meetings, and expectations, moments to touch base, to develop a plan.

In a partnership characterized by an incremental tempo, we improvised. We worked in and across the urban inequalities that divided us, navigating conflicts that so easily could have torn us apart. In moments of rupture, epiphanies of understanding were wrought, presumptions torn aside. On this foundation our rhythm was both slow and quick. We jumped to respond, to shift when things went slightly wrong. We tinkered, sometimes changed plans radically to make things work, to check and rework, to entwine and fuse.

Research teams eased into intimacies in their repetitions, in uneven, incremental jumps of confidence, in a steady building of collegiality. Off the bus by Gerty's house, with quicker greetings each week, clear on a plan, teams moved up and down neighborhood streets, in and out of homes. This process had a

Chapter Seven

meter, paced across the research process, its flow, its modulations. Its punctuation was marked in my insistence that we "turn assumptions into questions." The lingering goodbyes at the bus lasted longer each week. Picking up tempo, collectively and individually, we moved forward. This mode was the heart of our process, through which we researched, the mode through which partners and I taught together.

In this was the hard work of active listening to respond carefully to university notions of rigor, of some sense of social science truth, my partners' truths rooted in hard-fought experience, in struggle, in life in this city. In the partnership's back and forth, in these syncopated rhythms, these imperatives sat side by side, juxtaposed, a powerful mix, entwined in their contrast. The way we worked together was no longer method; it became substance, the inspiration for our questions and our practice. The sharing of the work was no longer at the university alone, it was in the neighborhood, inspired in the city, rooted in everyday critical urgent questions, in debates on justice, in rights, in claims to the city. The teaching was no longer my work alone, it was joint, rich, resonant, responsive, rooted in real life.

Across the slow time of a decade (or two), I changed form to write the rhythm of this partnership. In its stories and ordinary words, I found a vocabulary, a register of the partnership, its beat and dissonances, the tempos in and through which we worked. In this powerful, syncopated, sometimes discordant rhythm, we built theory in partnership: relational, embodied, experimental, sustained over a decade.

Theorizing in Partnership

This book has celebrated theorizing in partnership. This is a mode of theorizing the city otherwise: in its inspirations; in the practices of teaching and research that shaped it; in the publications through which it lives on; in its complexities, the compromises that made it functional, as well as meaningful; and its end. The partnership proved a collaborative problem space, a vehicle in which Gerty and I could build a process, a form, and a logic of researching and teaching collaboratively. We built an ethos to work and research together, to overcome and live with tensions. It shaped the logics of our research questions, of the conversations we encouraged and provoked, of the themes we pursued. Its intellectual logics sprang from movements for justice and rights and debates about access and knowledge.

In these fundamental tenets, we found a rhythm, a way to work and move, a practice. I found a way "to stay, not run," to return to the chairman's provo-

Theorizing the City Otherwise   **145**

cation with which I started this book. In this practice, I reimagined and refigured scholarship, rooted in context, transparent to its making, its travels, immersed in its collaborative form. Interlaced together, the partnership's content and form embodied a politics and poetics of collaborative knowledge production, resonant in the partnership's publics in and across the city.

The partnership decentered scholarly knowledge, making it one of many elements that constituted our collaboration. Through it we embraced partners' expertise and recalibrated the community—and neighborhoods like it across the city—as places of valid knowledge making. It shifted research, the analytical objects that came to view, the matters of concern and preoccupation, the commitments in which they were rooted, the registers through which they were made visible and gained traction and relevance. In this refiguring, we broadened the "intellectual" terrain and its theorists—us—partners and collaborators, inhabitants, urbanites. It opened understanding of the urban to multiple voices and arguments, to city publics, to diverse forms of knowledge and power. It opened it up to its high stakes, its politics, the urgency of multiple crises, the labor of everyday struggle, the politics of inequality that shape this city and cities around the world. It opened us up to its high hopes too, unearthed in the contingencies and surprises, in successes and achievements, accomplishments large and small, to the textures of life, its joys, passions, anxieties, and pain.

Urban theory in partnership is an imperative that urban theory open itself up to its making, to its practices, to its contexts, to thick forms of theory building.

At the heart of urban research are the myriad ways in which urbanists of all sorts speak, work, and learn with community leaders and residents, activists and policy makers, and the state. In our research and in our teaching, we interview, listen to, engage, argue with, and lobby experts who range from ordinary people to policy makers, activists, public intellectuals, artists, and politicians. Our questions are rooted in and are inspired by the soil of the city itself, its land divisions, its geopolitics of wealth and inequality, in varied city infrastructures, the way they work and fail, in city politics, its mobilizations and contentions. These debates preoccupy us, and our field. Collaborations of all sorts, even if not named as such, are an imperative, a condition of access, of entry into city sites and processes, a form of participation and association at the heart of urban work. They root urban work in expertise in the city and beyond the academy. Urban theorizing in partnership is an epistemological and political recalibration, which decenters the university and expands sites of

Chapter Seven

knowledge production and intellectual work across the city. In the relationalities at the heart of collaborative urban work we can be open to multiple voices and arguments, to city publics and diverse forms of knowledge and power.

Urban theorizing in partnership is an invitation to build collaborative practices and to write the varied stories that make them work. It is a call to put at the heart of our university practice and scholarly work that which matters to those with whom we make knowledge, those with and by whom we are inspired. It upends a presumption that scholarly research holds relevance for communities and the city in and of itself. It challenges simplistic notions that the university might extend itself, in its present form, to engage the city, to be "socially responsive." It is a rupture through which we can rethink university practices by embracing the collaborations that inspire our work. It is an invitation to form partnerships, liaisons of many sorts; to commit to them, to nurture them, to account to them, and to return to them.

IN MEMORY OF GERTRUDE SQUARE

On the twenty-first of August 2021 Aunty Gerty passed away. She was the backbone of her family, her community, and of my growth as an urbanist and academic. I dedicate this book to her.

> I hear your voice.
> Just, a trace of steel.
> Shoes repurposed
> Once
> Twice
> Thrice
> Homemade heels add inches.
> We squeeze them on.
> From these heady heights,
> In your steps,
> We walk on.

FIGURE 22. Gerty and Sophie, in the early years

BIBLIOGRAPHY

Askins, K. 2018. "Feminist geographies and participatory action research: Co-producing narratives with people and place." *Gender, Place and Culture,* 25,9: 1277–1294.

Autonomous Geographies Collective. 2010. "Beyond scholar activism: Making strategic interventions inside and outside the neoliberal university." *ACME: An International E-Journal for Critical Geographies,* 9,2: 245–275.

Ballard, R. 2015. "Geographies of development III: Militancy, insurgency, encroachment and development *by* the poor." *Progress in Human Geography,* 39,2: 214–224.

Bénit-Gbaffou, C., ed. 2015. *Popular Politics in South African Cities: Unpacking Community Participation.* Pretoria: Human Sciences Research Council Press.

Bénit-Gbaffou, C., S. Charlton, S. Didier, and K. Dörmann. 2019. *Politics and Community-Based Research: Perspectives from Yeoville Studio, Johannesburg.* Johannesburg: Wits University Press.

Bhan, G. 2019. "Notes on a Southern urban practice." *Environment and Urbanization,* 31,2: 639–654.

Bhan, G., S. Srinivas, and V. Watson, eds. 2018. *The Routledge Companion to Planning in the Global South.* London: Routledge.

Bruinders, S. 2006. "'This is our sport!' Christmas band competitions and the enactment of an ideal community." *SAMUS: South African Music Studies,* 26,1: 110–126.

Bruinders, S. 2010. "Parading respectability: The Christmas Bands Movement in the Western Cape, South Africa and the constitution of subjectivity." *African Music: Journal of the International Library of African Music,* 8,4: 69–83.

Bruinders, S. 2017. *Parading Respectability: The Cultural and Moral Aesthetics of the Christmas Bands Movement in the Western Cape, South Africa.* Grahamstown: African Humanities Programme.

Bunge, W. 2011. *Fitzgerald: Geography of a Revolution.* Athens: University of Georgia Press.

Chari, S., and H. Donner. 2010. "Ethnographies of activism: A critical introduction." *Cultural Dynamics,* 22: 75–85.

Charlton, S. 2009. "Housing for the nation, the city and the household: Competing rationalities as a constraint to reform?" *Development Southern Africa,* 26,2: 301–315.

Bibliography

Charlton, S. 2018. "Spanning the spectrum: Infrastructural experiences in South Africa's state housing programme." *International Development Planning Review*, 40,2: 97–120.

Choi, S., A. Selmeczi, and E. Strausz. 2020. *Critical Methods for the Study of World Politics: Creativity and Transformation*. London: Routledge.

Cirolia, L. R., T. Görgens, M. van Donk, W. Smit and S. Drimie, eds. 2017. *Upgrading Informal Settlements in South Africa: Pursuing a Partnership-Based Approach*. Cape Town: Juta.

City of Cape Town. 2020. *State of Cape Town Report 2020. Visual Summary of the Full Report*. Policy and Strategy Department: Cape Town. https://resource.capetown.gov.za/documentcentre/Documents/City%20research%20reports%20and%20review/State_of_Cape_Town_2020_Visual_Summary.pdf.

Comaroff, J., and J. L. Comaroff. 2012. *Theory from the South: Or, How Euro-America Is Evolving toward Africa*. London: Routledge.

Connell, R. 2007. *Southern Theory: The Global Dynamics of Knowledge in Social Science*. Malden, Mass.: Polity.

Cresswell, T. 2021. "Beyond geopoetics: For hybrid texts." *Dialogues in Human Geography*, 11,1: 36–39.

Crush, J., and A. Chikanda, eds. 2015. *Mean Streets: Migration, Xenophobia and Informality in South Africa*. Waterloo: Southern African Migration Programme.

Dasgupta, S., and N. Wahby. 2021. "Introduction: Beyond a standardised urban lexicon: Which vocabulary matters?" *International Development Planning Review*, 43,4: 419–433.

Dauphinee, E. 2010. "The ethics of autoethnography." *Review of International Studies*, 36,3: 799–818.

Dauphinee, E. 2013a. *The Politics of Exile*. London: Routledge.

Dauphinee, E. 2013b. "Writing as hope: Reflections on *The Politics of Exile*." *Security Dialogue*, 44,4: 347–361.

Dickens, L., and T. Edensor. 2022. "Dreamlands: Stories of enchantment and excess in a search for lost sensations." *Cultural Geographies*, 29,1: 23–43.

Dierwechter, Y. 2004. "Dreams, bricks, and bodies: Mapping 'neglected spatialities' in African Cape Town." *Environment and Planning A: Economy and Space*, 36,6: 959–981.

Favish, J., and J. McMillan. 2009. "The university and social responsiveness in the curriculum: A new form of scholarship." *London Review of Education*, 7,2: 169–179.

Fuller, D. 2008. "Public geographies: Taking stock." *Progress in Human Geography*, 32,6: 1–11.

Gaulier, A., and D. C. Martin. 2017. *Cape Town Harmonies: Memory, Humour and Resilience*. Cape Town: African Minds.

Huchzermeyer, M. 2001. "Housing for the poor? Negotiated housing policy in South Africa." *Habitat International*, 25,3: 303–331.

Inayatullah, N., ed. 2011. *Autobiographical International Relations: I, IR*. London: Routledge.

Inayatullah, N. 2013. "Pulling threads: Intimate systematicity in *The Politics of Exile*." *Security Dialogue*, 44,4: 319–348.

Bibliography   **151**

James, D. 2012. "Money-go-round: Personal economies of wealth, aspiration and indebtedness." *Africa*, 82,1: 20–40.

Jeppie, S. 1990. "Popular culture and carnival in Cape Town: The 1940s and 1950s." In *The Struggle for District Six Past and Present*, edited by S. Jeppie and C. Soudien, 67–87. Cape Town: Buchu Books.

Katz, C. 2017. "Revisiting minor theory." *Environment and Planning D: Society and Space*, 35,4: 596–599.

Kindon, S., and S. Elwood. 2009. "Introduction: More than methods—Reflections on participatory action research in geographic teaching, learning and research." *Journal of Geography in Higher Education*, 33,1: 19–32.

Lalu, P. 2012. "Still searching for the 'human.'" *Social Dynamics*, 38,1: 3–7.

Landau, L. B. 2012. *Exorcising the Demons Within: Xenophobia, Violence and Statecraft in Contemporary South Africa*. Johannesburg: Wits University Press.

Leitner, H., J. Peck, and E. Sheppard. 2007. *Contesting Neoliberalism: Urban Frontiers*. London: Guilford Press.

Lemanski, C., and S. Oldfield. 2009. "The parallel claims of gated communities and land invasions in a Southern city: Polarised state responses." *Environment and Planning A*, 41,3: 634–648.

Levenson, Z. 2018. "The road to TRAs is paved with good intentions: Dispossession through delivery in post-apartheid Cape Town." *Urban Studies*, 55,14: 3218–3233.

Levenson, Z. 2021. "Becoming a population: Seeing the state, being seen by the state, and the politics of eviction in Cape Town." *Qualitative Sociology*, 367–383.

Levenson, Z. 2022. *Delivery as Dispossession: Land Occupation and Eviction in the Post-Apartheid City*. Oxford: Oxford University Press.

Lorimer, H., and H. Parr. 2014. "Excursions—Telling stories and journeys." *Cultural Geographies*, 21,4: 543–547.

Mabin, A. 1984. "WEJGE: The genesis of an exploration in geographical learning." *South African Geographer*, 12,1: 69–79.

Martin, D. 1999. *Coon Carnival: New Year in Cape Town, Past to Present*. Cape Town: David Philip.

Mason, W. 2021. "On staying: Extended temporalities, relationships and practices in community engaged scholarship." *Qualitative Research*, 0,0:1–20.

McFarlane, C. 2011. *Learning the City: Knowledge and Translocal Assemblage*. Hoboken, N.J.: John Wiley & Sons.

Miller, C. 2007. "'Julle kan ma New York toe gaan, ek bly in die Manenberg': An oral history of jazz in Cape Town from the mid-1950s to the mid-1970s." In *Imagining the City: Memories and Cultures in Cape Town*, edited by S. Field, R. Meyer, and F. Swanson, 133–149. Cape Town: Human Sciences Research Council Press.

Millstein, M. 2020. "'If I had my house, I'd feel free': Housing and the (re)productions of citizenship in Cape Town, South Africa." *Urban Forum*, 31,3: 289–309.

Miraftab, F. 2004. "Invited and invented spaces of participation: Neoliberal citizenship and feminists' expanded notion of politics." *Wagadu*, 1,1: 1–7.

Miraftab, F. 2009. "Insurgent planning: Situating radical planning in the global south." *Planning Theory*, 8,1: 32–50.

Bibliography

Mitchell, K. 2008. *Practising Public Scholarship: Experiences and Possibilities beyond the Academy.* Oxford: Wiley-Blackwell.

Mrs Kinpaisby. 2008. "Taking stock of participatory geographies: Envisioning the communiversity." *Transactions of the Institute of British Geographers,* 33,3: 292–299.

Nagar, R. 2002. "Footloose researchers, 'traveling' theories, and the politics of transnational feminist praxis." *Gender, Place and Culture,* 9,2: 179–186.

Nagar, R. 2012. "Storytelling and co-authorship in feminist alliance work: Reflections from a journey." *Gender, Place and Culture,* 20,1: 1–18.

Nagar, R. 2014. *Muddying the Waters: Coauthoring Feminisms across Scholarship and Activism.* Champaign: University of Illinois Press.

Nagar, R. 2019. *Hungry Translations: Relearning the World through Radical Vulnerability.* Champaign: University of Illinois Press.

Ngwenya, N., and L. R. Cirolia. 2021. "Conflicts between *and* within: The 'conflicting rationalities' of informal occupation in South Africa." *Planning Theory and Practice,* 22,1: 1–16.

Nyamnjoh, F. 2012. "Potted plants in green houses: A critical reflection on the resilience of colonial education." *Journal of Asian and African Studies,* 47,2: 129–154.

Oldfield, S. 2008a. "Who's serving whom? Partners, processes and products in service-learning projects in South Africa." *Journal of Geography in Higher Education,* 32,2: 269–285.

Oldfield, S. 2008b. "Teaching through service-learning projects in urban geography courses at the University of Cape Town." In *Service-Learning in the Disciplines: Lessons from the Field.* Higher Education Quality Committee / JET Education Services. Pretoria: Council on Higher Education.

Oldfield, S. 2015. "Between the academy and activism: The urban as political terrain." *Urban Studies,* 52,11: 2072–2086.

Oldfield, S., and S. Greyling. 2015. "Waiting for the state: A politics of housing in South Africa." *Environment and Planning A,* 47,5: 1100–1112.

Oldfield, S., S. Parnell, and A. Mabin. 2004. "Engagement and reconstruction in critical research: Negotiating urban practice, policy and theory in South Africa." *Journal of Social and Cultural Geography,* 5,2: 285–300.

Oldfield, S., and Z. Patel. 2016. "Engaging geographies: Negotiating positionality and building relevance." *South African Geographical Journal,* 98,3: 505–514.

Oldfield, S., and K. Stokke. 2006. "Building unity in diversity: Social movement activism in the Western Cape Anti-Eviction Campaign." In *Globalisation, Marginalisation and New Social Movements,* edited by A. Habib, I. Valodia, and R. Ballard, 25–49. Durban: University of KwaZulu-Natal Press.

Oswin, N., and G. Pratt. 2021. "Critical urban theory in the 'Urban Age': Ruptures, tensions, and messy solidarities." *International Journal of Urban and Regional Research,* 45,4: 585–596.

Parnell, S. 2007. "The academic-policy interface in post-apartheid urban research: Personal reflections." *South African Geographical Journal,* 89,2: 111–120.

Peake, L. 2016. "The twenty-first century quest for feminism and the global urban." *International Journal of Urban and Regional Research,* 40,1: 219–227.

Pieterse, E. 2014. "Epistemological practices of southern urbanism." Paper presented for the African Centre for Cities Seminar Series. Cape Town: African Centre for Cities, University of Cape Town. https://www.africancentreforcities.net/wp-content/uploads/2014/02/Epistemic-practices-of-southern-urbanism-Feb-2014.pdf.

Pratt, G. 2012. *Families Apart: Migrant Mothers and the Conflicts of Labor and Love.* Minneapolis: University of Minnesota Press.

Richardson, L. 2020. "Writing: A method of inquiry." In *Handbook of Qualitative Research,* edited by N. Denzin and Y. Lincoln, 923–948. Thousand Oaks, Calif.: Sage.

Robinson, J., and A. Roy. 2016. "Debate on global urbanisms and the nature of urban theory." *International Journal of Urban and Regional Research,* 40,1: 181–186.

Rogerson, C. M. 1996. "Urban poverty and the informal economy in South Africa's economic heartland." *Environment and Urbanization,* 8,1: 167–179.

Routledge, P., and K. D. Derickson. 2015. "Situated solidarities and the practice of scholar-activism." *Environment and Planning D: Society and Space,* 33,3: 391–407.

Roy, A. 2020. "'The shadow of her wings': Respectability politics and the self-narration of geography." *Dialogues in Human Geography,* 10,1: 19–22.

Salo, E. 2004. "Negotiating gender and personhood in the new South Africa: Adolescent women and gangsters in Manenberg township on the Cape Flats." *European Journal of Cultural Studies,* 6,3: 345–365.

Salo, E. 2018. *Respectable Mothers, Tough Men and Good Daughters: Producing Persons in Manenberg Township.* Bamenda: Langaa Research and Publishing Common Initiative Group.

Sandercock, L. 2003. "Out of the closet: The importance of stories and storytelling in planning practice." *Planning Theory and Practice,* 4,1: 11–28.

Sangtin Writers and R. Nagar. 2006. *Playing with Fire: Feminist Thought and Activism through Seven Lives in India.* Minneapolis: University of Minnesota Press.

Saville, S. M. 2021. "Towards humble geographies." *Area,* 53,1: 97–105.

Scott, D. 2014. "The temporality of generations: Dialogue, tradition, criticism." *New Literary History,* 45,2: 157–181.

Seekings, J. 2011. "The changing faces of urban civic organisation." *Transformation: Critical Perspectives on Southern Africa,* 75,1: 140–161.

Selmeczi, A. 2012. "Abahlali's vocal politics of proximity: Speaking, suffering and political subjectivization." *Journal of Asian and African Studies,* 47,5: 498–515.

Selmeczi, A. 2014. "Dis/placing political illiteracy: The politics of intellectual equality in a South African shack dwellers' movement." *Interface: Journal for and about Social Movements,* 6,1: 230–265.

Shannon, J., K. B. Hankins, T. Shelton, A. J. Bosse, D. Scott, D. Block, H. Fischer, L. E. Eaves, J. Jung, J. Robinson, P. Solís, H. Pearsall, A. Rees and A. Nicolas. 2020. "Community geography: Toward a disciplinary framework." *Progress in Human Geography,* 45,5: 1147–1168.

Sitas, A. 2004. *Voices That Reason: Theoretical Parables.* Leiden: Brill Academic Publishers.

Skinner, C. 2008. "The struggle for the streets: Processes of exclusion and inclusion of street traders in Durban, South Africa." *Development Southern Africa,* 25,2: 227–242.

Swarr, A., and R. Nagar. 2010. *Critical Transnational Feminist Praxis.* Albany: State University of New York Press.

Bibliography

Tuck, E. 2009. "Suspending damage: A letter to communities." *Harvard Educational Review*, 79,3: 409–428.

Tuck, E., M. Smith, A. M. Guess, T. Benjamin, and B. K. Jones. 2014. "Geotheorizing Black/land: Contestations and contingent collaborations." *Departures in Critical Qualitative Research*, 3,1: 52–74.

Tuck, E., and K. W. Yang. 2012. "Decolonization is not a metaphor." *Decolonization: Indigeneity, Education & Society*, 1,1: 1–40.

Tuck, E., and K. W. Yang. 2014. "R-words: Refusing research." In *Humanizing Research: Decolonizing Qualitative Inquiry with Youth and Communities*, edited by D. Paris and M. T. Winn, 223–248. Thousand Oaks, Calif.: Sage.

Visagie, J., and I. Turok. 2021. "Driven further apart by the pandemic? Contrasting impacts of COVID-19 on people and places." National Income Dynamics Study (NIDS)—Coronavirus Rapid Mobile Survey (CRAM). https://cramsurvey.org/wp-content/uploads/2021/07/12.-Turok-I.-_-Visagie-J.-2021-Drive-apart_-Contrasting-impacts-of-COVID-19-on-people-and-places.pdf.

Winkler, T. 2013. "At the coalface: Community-university engagements and planning education." *Journal of Planning Education and Research*, 33,2: 215–227.

INDEX

Abrahams, Nas Abdul, 47, 63–65
academic practices, 5, 11, 14, 27, 60, 65. *See also* university context
academic writing, 9, 19–20, 116–119, 138. *See also* publications
acknowledgment, 112
activism, 18, 25, 30, 44–47, 60, 71, 79
admiration, 94, 103–106
African National Congress (ANC), 30
African Zonks, 50
Agste Laan, 42–44, 61–65, 71, 112–115, 136
Ahmed, Sara, 17
alcohol, 126
ANC (African National Congress), 30
anger, 18, 66–68
anonymity in review process, 118
anti-eviction campaign. *See* Western Cape Anti-Eviction Campaign
apartheid, 18, 27, 34, 36, 46, 50–51, 130
archive of partnership, 7–8, 21, 27, 132–135, 142
Arendse, Washiela, 14, 27, 75, 132–133
articles, journal, 116–119
assessments of students, 101–106
assumptions, 19, 71, 89–95, 144
Aziza, 67, 95

babysitting, 74
backyarders project, 15, 26, 31, 34–38, 70–71
banners and boards of troupes, judging of, 66–68
base maps, 34, 43–44, 53, 62–63. *See also* maps
Bill of Rights, 37
boards and banners of troupes, judging of, 66–68

book chapters, 116–119
books, short: *My 7de Laan*, 42, 107, 123, 128–130, 134; *The Valhalla Community Entertainers*, 52, 107, 122–126, 130, 134
bungalows, 31, 36, 95–96. *See also* backyarders project
businesses, 46, 52–56, 71–74, 100, 108, 119–122, 130, 136. See also *Yellow Pages*

Cape Carnival on Tweede NuweJaar. *See* minstrels
Cartesian god's-eye view, 113
cell phones, 35
children, 74, 123
city government (City of Cape Town): Agste Laan and, 43; Civic and, 15, 30–33, 95; code of conduct, 71; court case, 38–42; housing policies, 37; minstrels and, 126; public housing project, planned, 39, 58, 76; Sewende Laan and, 38–42, 126–127; unemployment, 53
Civic: Agste Laan project, 42–43, 63–64; backyarders project, 35–36; businesses project, 72–73; gangsterism and, 75–77; introduction to, 27–34; minstrels project, 47–52, 68, 123, 126, 130; partnership with, 4, 6, 14–16, 25–26, 56–60, 75, 77, 80, 86, 132, 136, 143; research project on, 16, 44–47; Sewende Laan project, 38–42, 123, 126–127; water, access to, 68–70
code of conduct, City of Cape Town, 71
collaboration, 3–5, 9, 14–18, 22–23, 61, 84, 89, 136–138
colonialism, 11, 18, 50, 61

156 Index

"coloured" classification under apartheid, 27, 31
"coloured" identity, 47, 50–51, 126
community development, 27, 47–52, 126, 139
Community Research Group, 15, 34
community work, 44–47, 139
compensation. *See* employment
complaint, 74–75
compromise, 17, 60, 65, 75, 139
conflict. *See* tensions
consent, 87–88, 90, 91, 142
contradictions, 8, 22, 60–61, 71, 73, 79, 108, 139
councilors, 47, 63–65, 71, 123
court case, Sewende Laan, 38–42, 127
Covid-19 pandemic, 53
crime, 75–77
crises, 57

Dan, Uncle, 6, 36, 70–71, 127, 130
Dauphinee, Elizabeth, 18–19, 20; *The Politics of Exile*, 19
decolonialization, 11
dignity, 96, 138
directory of neighborhood businesses. See *Yellow Pages*
District Six, 27, 31
"doing words," 141–144
drug trade, 73, 76–77, 126

emotions, 69, 137
employment, 16, 53, 63–64, 74–75, 138
"engineering work," 50, 123
Environment and Planning A, 116
essentialism, 102
events. *See* parties
evictions, 14–15, 27, 30, 39, 56, 116. *See also* forced removals
"exhibits," 87

Fadielah, Aunty, 63, 64, 78, 87
family, 36, 119, 139
Fatima, 6, 69, 78, 133
fear, 71–74
field notes, 21, 81, 96, 97, 132
fieldwork, 31, 34, 82–83, 85, 89
final papers, 81, 96, 101–106
financial compensation. *See* employment
fires, 57
Firm, the, 76–77
food, in fridge, 79
forced removals, 27, 30, 31, 51. *See also* evictions

"foreigners," 52–53, 71–74, 122
fridge, empty, 79
friendships, 57, 59
funding of projects, 17, 35

gangsterism, 52, 73, 76–78, 84–88, 102, 122, 126
gardens, 69
genres of publication. *See* publications
global south, 24
god's-eye view, 113
grants. *See* social welfare grants
guidelines for research, 88–91

"high risk area," Valhalla Park as, 93
home-based businesses. *See* businesses
homelessness, 31, 38–42
homes, 39, 42, 95–96, 113
housing, 10, 37–42, 77
humiliation, 69

"I," 22
Inayatullah, Naeem, 18, 19, 20
inequality, 11, 16, 18, 53, 79, 92–95
informal trading, 46, 52–53, 72, 100, 119, 122, 130. *See also* businesses
interviews, 34, 89, 90, 91, 94–97
invisibility of backyards, 34
invitations and refusals, 61, 79

James, D., 100
journal articles, 116–119
journals, weekly, 91–96
judging of minstrel competition, 50, 66–68

Kaaps (language), 138
Kaapse Klopse. *See* minstrels
Klopse. *See* minstrels
Klopse Board, 52, 66, 123, 126
knowledge production, 4, 11, 21, 27, 135, 145
Koekie, 6, 15, 75

landlords, 36
land occupation, 16, 31, 37–44, 64, 71
laws, 69–70
leaders and leadership, 27–30, 64–65
learning, 27, 81–84, 89, 90, 91–92, 94–95, 101, 106
Leaticia, 74, 79
Lefien, 75, 108, 127, 130
Legal Resources Centre (LRC), 39, 127, 134

Index   **157**

legitimacies, 64–65
Leitner, Helga, 116
Lemanski, Charlotte, 116
letters from partners to students, 140
Lissie, Aunty, 39, 42
listening, 96–97, 144
Loggenberg, Oscar, 77–78
logistical planning, 35
love, 101–106
LRC. *See* Legal Resources Centre

"making ends meet" project. *See* businesses
Malay Choirs, 51
maps, 34–37, 40–44, 53, 62–63, 85, 107–108, 112–115, 134
Masnoena, 61, 63, 64–65, 75, 133
Mason, W., 27
McFarlane, Colin, 27
methamphetamine (*tik*) addiction, 77, 126
Mina, 6, 74–75, 95, 127, 130
minstrels, 30, 47–52, 66–68, 122–126, 130, 134, 136
Minstrels Board. *See* Klopse Board
municipal services. *See* services, provision of
murders, 77–78. *See also* violence
My 7de Laan, 42, 107, 123, 126–130, 134

Nagar, Richa, 17, 20
Naomi, 6, 75
narrative writing, 3–5, 7–9, 18–25
National Income Dynamics Study, 53
Neoliberal Futures, 117
neoliberalism, 116–117
New Crossroads, 14, 31
newspaper clippings, 32–33, 132
"not once, not twice, but thrice," 30, 139

ordinary people, 4–5, 11, 24, 84, 96, 113, 137
ordinary words, 23–25, 138–144
orientation routine, 80, 86–88
"otherwise," doing urban studies, 7–9, 11, 20, 24, 27, 144

participatory practice, 27, 46, 109–112, 133
parties, 38, 42, 73, 81, 109, 140
partners, 28–29, 70–71, 75, 82–83, 89, 94–95, 101–106, 130–132, 140
partnership: archive of, 7–8, 21, 27, 132–135, 142; beginning of, 26; characters of, 5–7; contradictions of, 60; end of, 58, 77–79;

leadership of, 74; ordinary words, 139–144; practices of, 21–24, 117–118; rhythms of, 44, 56–59, 136–138, 143–144; value of, 3–5, 9, 14–18, 20–21, 25, 144–146
pedagogy, 25, 84, 106
peer-review process, 117
photographs: of Agste Laan, 43, 112–115; of businesses, 54–55; of Gerty and Sophie, 147; of minstrels, 48–49; of partners, 28–29; on posters, 109–111; of posters event, 133; of researchers, 82–83, 89; of Sewende Laan, 40–41; taking of, 87–88, 90, 91; of Valhalla Park, 12–13
Pieterse, Edgar, 11
police, 39, 57–58, 66, 68, 93
politicians, 126
politics, 11, 53, 58–59, 79, 84, 91, 94–95, 117–119, 137
Politics of Exile, The (Dauphinee), 19
posters, 42, 51, 81, 107–112, 133–134
postgraduate students and teaching, 81, 91–92, 131–132
Pratt, G., 17
preparatory work for projects, 62, 86
pride, 51, 123, 139
problem spaces, 17–18
projects, 15–16, 26, 56–58, 77–78; research, on Civic, 16, 44–47. *See also* Agste Laan; backyarders project; businesses; minstrels; Sewende Laan
publications, 16, 97–101, 107–108, 132–135, 139–140, 142; book chapters and journal articles, 116–119; weekly journals, 91–96. *See also* books, short; field notes; maps; posters; reports; *Yellow Pages*
public housing, 30, 37, 44, 131
public housing project, planned, 16, 39, 58, 76
public participation. *See* participatory practice

qualitative analyses, 91, 100
"questioning what we know," 19, 89–95

racialized divide, 53, 126
racial stereotypes, 130
racism, 102, 126
Raksha, 7, 140
recognition, 112, 140
refiguring, 17
refusals and invitations, 61, 79
relationships, 36, 57, 90, 140

Index

rental housing, 30–31, 36–37, 44, 131
reports, 36, 98–99, 101–106, 132–133
research: approaches to, 15–16, 21, 90, 135; outcomes, 142; "questioning what we know," 91–95; rhythm of, 56–57; "spin off," 130–132; "staying, not running," 10–11; value of, 31
researchers, 10, 71, 82–83, 112–113, 118
research literature, 95–97
research parties. *See* parties
research projects. *See* projects
research publications. *See* publications
respect, 88–91, 103, 106
review process, academic, 117–119
rhythms of partnership, 56–58, 136, 143–144
Richardson, L., 20
Rogerson, C. M., 100
roles, multiple and shifting, 6–7, 68, 71, 75, 89, 140–141
Rosemary, 6, 75, 102
Rosenberg, George, 30, 46, 116, 127

Salo, E., 20
Sandercock, Leonie, 20
Saskia, 7, 131
schedules, 88–89
scholarly writing. *See* academic writing
Seekings, J., 46
Selmeczi, A., 20
sensitivity, 90
services, provision of, 26, 37, 39, 43; access to water, 68–70, 111
Sewende Laan, 38–42, 71, 113, 123, 126–130, 134, 136
Sewende Laan Committee, 71
"shacks," 31, 95–96. *See also* backyarders project
Sheppard, Eric, 116
Shireen, 6, 140
shops. *See* businesses
short books. *See* books, short
Siân, 7, 131
Sitas, Ari, 11
slavery, 50, 51, 126
slowness of approach, 56, 143
small businesses. *See* businesses
social welfare grants, 72, 102, 130
Somali traders, 71, 72–74, 122
South African Constitutional Bill of Rights, 37
South African Police. *See* police
"southern theory," 24

"spin off" research, 130–132
Square, Gertrude (Gerty): Agste Laan project, 43; backyarders project, 35; businesses research project, 73, 119; Civic, research project on, 44; dedication to, 147; leadership of, 61–67, 127, 147; meets author, 27, 30; minstrels project, 50, 52, 66–68, 122, 126; partnership with, 5–6, 8, 14–15, 25, 56–57, 74–75, 86–88, 132, 143–144; photograph of, 147; Sewende Laan project, 26, 34, 127; water, access to, 69–70
Stanfield, Colin, 76
state-built housing. *See* public housing
state failures, 46
"staying, not running," 10–14, 25, 27, 144–145
stigma, 25, 36, 52, 126, 139
Stokke, Kristian, 14, 116–117
storytelling, 21–23, 97, 100, 136–138. *See also* narrative writing
street trading, 53. *See also* businesses
students: assessment of, 101–106; contradictions, 60; first project with, 35–37; orientation routine, 80–88; partners and, 50, 56–57, 69–71, 101–106, 140; partnership, 6–7; posters, 108–109; "questioning what we know," 91–94; research literature, 95–96; "spin off" research, 130–132
Suki, 6, 130
Swarr, Amanda, 17

"taking another look," 102–103
teaching, 8–9, 25, 56–57, 80–84, 106, 118, 140–142
tensions, 43, 52–53, 57, 60–65, 69–76, 79
"Territorialization in Cities of the South," 116
theorizing, 24–25, 84, 100–101, 143–146
tik (methamphetamine) addiction, 77, 126
Top Copy printers, 122
"township revolt," 46
trompoppies, 123
trust, 18, 56–57, 67, 74, 140, 143
"truth-telling," 138
Tuck, E., 61

unemployment, 16, 53. *See also* employment
United Front Civic Organisation. *See* Civic
university context, 3, 10–18, 25, 27, 60, 80–84, 89, 106, 137–138, 143–146. *See also* academic practices

Urban Futures Conference (2000), 14
urbanism, 5, 24
urban theorizing, 24–25, 84, 100–101, 143–146

Valhalla Community Entertainers, The, 52, 107, 122–126, 130, 134
Valhalla Park: history of, 31; introduction to, 12–13; map of, 85
Valhalla Park minstrel troupe, 47–52, 66, 122–126
Valhalla Park United Front Civic Organisation. *See* Civic
verbs, 141–144
violence, 52, 57–58, 60–61, 71–77, 102, 119, 122, 139
vocabulary (ordinary words), 23–25, 138–144

water, access to, 68–70, 111. *See also* services, provision of
"we," 22

weekly journals, 91–96
Wendy houses, 31, 36. *See also* backyarders project
Western Cape Anti-Eviction Campaign, 14–15, 27, 30, 116
women in minstrel troupes, 51
writing, academic. *See* academic writing
writing practices, 16, 97–101, 107–108, 132–135, 139–140, 142; book chapters and journal articles, 116–119; weekly journals, 91–96. *See also* books, short; field notes; maps; posters; reports; *Yellow Pages*

xenophobia, 52–53, 56, 71–74, 119, 122

Yellow Pages, 56, 108, 119–122, 133–134

Zaaida, 6, 51, 68, 70, 78, 133, 140
"zoo," 87

GEOGRAPHIES OF JUSTICE AND SOCIAL TRANSFORMATION

1. *Social Justice and the City*, rev. ed.
   BY DAVID HARVEY

2. *Begging as a Path to Progress: Indigenous Women and Children and the Struggle for Ecuador's Urban Spaces*
   BY KATE SWANSON

3. *Making the San Fernando Valley: Rural Landscapes, Urban Development, and White Privilege*
   BY LAURA R. BARRACLOUGH

4. *Company Towns in the Americas: Landscape, Power, and Working-Class Communities*
   EDITED BY OLIVER J. DINIUS AND ANGELA VERGARA

5. *Tremé: Race and Place in a New Orleans Neighborhood*
   BY MICHAEL E. CRUTCHER JR.

6. *Bloomberg's New York: Class and Governance in the Luxury City*
   BY JULIAN BRASH

7. *Roppongi Crossing: The Demise of a Tokyo Nightclub District and the Reshaping of a Global City*
   BY ROMAN ADRIAN CYBRIWSKY

8. *Fitzgerald: Geography of a Revolution*
   BY WILLIAM BUNGE

9. *Accumulating Insecurity: Violence and Dispossession in the Making of Everyday Life*
   EDITED BY SHELLEY FELDMAN, CHARLES GEISLER, AND GAYATRI A. MENON

10. *They Saved the Crops: Labor, Landscape, and the Struggle over Industrial Farming in Bracero-Era California*
    BY DON MITCHELL

11. *Faith Based: Religious Neoliberalism and the Politics of Welfare in the United States*
    BY JASON HACKWORTH

12. *Fields and Streams: Stream Restoration, Neoliberalism, and the Future of Environmental Science*
    BY REBECCA LAVE

13. *Black, White, and Green: Farmers Markets, Race, and the Green Economy*
    BY ALISON HOPE ALKON

14. *Beyond Walls and Cages: Prisons, Borders, and Global Crisis*
    EDITED BY JENNA M. LOYD, MATT MITCHELSON, AND ANDREW BURRIDGE

15. *Silent Violence: Food, Famine, and Peasantry in Northern Nigeria*
    BY MICHAEL J. WATTS

16. *Development, Security, and Aid: Geopolitics and Geoeconomics at the U.S. Agency for International Development*
    BY JAMEY ESSEX

17. *Properties of Violence: Law and Land-Grant Struggle in Northern New Mexico*
    BY DAVID CORREIA

18. *Geographical Diversions: Tibetan Trade, Global Transactions*
    BY TINA HARRIS

19. *The Politics of the Encounter: Urban Theory and Protest under Planetary Urbanization*
    BY ANDY MERRIFIELD

20. *Rethinking the South African Crisis: Nationalism, Populism, Hegemony*
    BY GILLIAN HART

21. *The Empires' Edge: Militarization, Resistance, and Transcending Hegemony in the Pacific*
    BY SASHA DAVIS

22. *Pain, Pride, and Politics: Social Movement Activism and the Sri Lankan Tamil Diaspora in Canada*
    BY AMARNATH AMARASINGAM

23. *Selling the Serengeti: The Cultural Politics of Safari Tourism*
    BY BENJAMIN GARDNER

24. *Territories of Poverty: Rethinking North and South*
    EDITED BY ANANYA ROY AND EMMA SHAW CRANE

25. *Precarious Worlds: Contested Geographies of Social Reproduction*
    EDITED BY KATIE MEEHAN AND KENDRA STRAUSS

26. *Spaces of Danger: Culture and Power in the Everyday*
    EDITED BY HEATHER MERRILL AND LISA M. HOFFMAN

27. *Shadows of a Sunbelt City: The Environment, Racism, and the Knowledge Economy in Austin*
BY ELIOT M. TRETTER

28. *Beyond the Kale: Urban Agriculture and Social Justice Activism in New York City*
BY KRISTIN REYNOLDS AND NEVIN COHEN

29. *Calculating Property Relations: Chicago's Wartime Industrial Mobilization, 1940–1950*
BY ROBERT LEWIS

30. *In the Public's Interest: Evictions, Citizenship, and Inequality in Contemporary Delhi*
BY GAUTAM BHAN

31. *The Carpetbaggers of Kabul and Other American-Afghan Entanglements: Intimate Development, Geopolitics, and the Currency of Gender and Grief*
BY JENNIFER L. FLURI AND RACHEL LEHR

32. *Masculinities and Markets: Raced and Gendered Urban Politics in Milwaukee*
BY BRENDA PARKER

33. *We Want Land to Live: Making Political Space for Food Sovereignty*
BY AMY TRAUGER

34. *The Long War: CENTCOM, Grand Strategy, and Global Security*
BY JOHN MORRISSEY

35. *Development Drowned and Reborn: The Blues and Bourbon Restorations in Post-Katrina New Orleans*
BY CLYDE WOODS
EDITED BY JORDAN T. CAMP AND LAURA PULIDO

36. *The Priority of Injustice: Locating Democracy in Critical Theory*
BY CLIVE BARNETT

37. *Spaces of Capital / Spaces of Resistance: Mexico and the Global Political Economy*
BY CHRIS HESKETH

38. *Revolting New York: How 400 Years of Riot, Rebellion, Uprising, and Revolution Shaped a City*
GENERAL EDITORS: NEIL SMITH AND DON MITCHELL
EDITORS: ERIN SIODMAK, JENJOY ROYBAL, MARNIE BRADY, AND BRENDAN O'MALLEY

39. *Relational Poverty Politics: Forms, Struggles, and Possibilities*
EDITED BY VICTORIA LAWSON AND SARAH ELWOOD

40. *Rights in Transit: Public Transportation and the Right to the City in California's East Bay*
BY KAFUI ABLODE ATTOH

41. *Open Borders: In Defense of Free Movement*
EDITED BY REECE JONES

42. *Subaltern Geographies*
EDITED BY TARIQ JAZEEL AND STEPHEN LEGG

43. *Detain and Deport: The Chaotic U.S. Immigration Enforcement Regime*
BY NANCY HIEMSTRA

44. *Global City Futures: Desire and Development in Singapore*
BY NATALIE OSWIN

45. *Public Los Angeles: A Private City's Activist Futures*
BY DON PARSON
EDITED BY ROGER KEIL AND JUDY BRANFMAN

46. *America's Johannesburg: Industrialization and Racial Transformation in Birmingham*
BY BOBBY M. WILSON

47. *Mean Streets: Homelessness, Public Space, and the Limits of Capital*
BY DON MITCHELL

48. *Islands and Oceans: Reimagining Sovereignty and Social Change*
BY SASHA DAVIS

49. *Social Reproduction and the City: Welfare Reform, Child Care, and Resistance in Neoliberal New York*
BY SIMON BLACK

50. *Freedom Is a Place: The Struggle for Sovereignty in Palestine*
BY RON J. SMITH

51. *Loisaida as Urban Laboratory: Puerto Rico Community Activism in New York*
BY TIMO SCHRADER

52. *Transecting Securityscapes: Dispatches from Cambodia, Iraq, and Mozambique*
BY TILL F. PAASCHE AND JAMES D. SIDAWAY

53. *Nonperforming Loans, Nonperforming People: Life and Struggle with Mortgage Debt in Spain*
BY MELISSA GARCÍA-LAMARCA

54. *Disturbing Development in the Jim Crow South*
BY MONA DOMOSH

55. *Famine in Cambodia: Geopolitics, Biopolitics, Necropolitics*
BY JAMES A. TYNER

56. *Well-Intentioned Whiteness: Green Urban Development and Black Resistance in Kansas City*
BY CHHAYA KOLAVALLI

57. *Urban Climate Justice: Theory, Praxis, Resistance*
EDITED BY JENNIFER L. RICE, JOSHUA LONG, AND ANTHONY LEVENDA

58. *Abolishing Poverty: Toward Pluriverse Futures and Politics*
BY VICTORIA LAWSON, SARAH ELWOOD, MICHELLE DAIGLE, YOLANDA GONZÁLEZ MENDOZA, ANA P. GUTIÉRREZ GARZA, JUAN HERRERA, ELLEN KOHL, JOVAN LEWIS, AARON MALLORY, PRISCILLA McCUTCHEON, MARGARET MARIETTA RAMÍREZ, AND CHANDAN REDDY

59. *Outlaw Capital: Everyday Illegalities and the Making of Uneven Development*
BY JENNIFER L. TUCKER

60. *High Stakes, High Hopes: Urban Theorizing in Partnership*
BY SOPHIE OLDFIELD

Printed in the USA
CPSIA information can be obtained
at www.ICGtesting.com
LVHW041922061023
760214LV00002B/35